Chris Defonseka
**Processing of Polymers**

# Also of interest

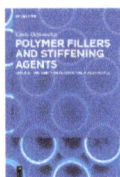

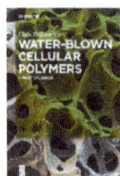

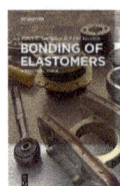

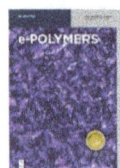

Chris Defonseka

# Processing of Polymers

—

DE GRUYTER

**Author**
Chris Defonseka
Toronto
Canada
defonsekachris@rogers.com

ISBN 978-3-11-065611-4
e-ISBN (PDF) 978-3-11-065615-2
e-ISBN (EPUB) 978-3-11-065642-8

**Library of Congress Control Number: 2020939673**

**Bibliographic information published by the Deutsche Nationalbibliothek**
The Deutsche Nationalbibliothek lists this publication in the Deutsche Nationalbibliografie;
detailed bibliographic data are available on the Internet at http://dnb.dnb.de.

# Preface

For many years now, the word plastic has been deeply ingrained into our society and forms one of the essential materials in daily life. Plastics are based on a wide spectrum of polymers, which have seen spectacular growth over the past many years. Polymers are generally derived from natural or synthetic sources and their versatility lies in their ability to be processed as single polymers, polymer blends or composites to produce materials that can be shaped and molded into different products with properties to suit end applications.

Polymers are materials composed of molecules of high molecular weights. The properties of these basic materials can be greatly enhanced to suit most applications by the incorporation of suitable additives. The uniqueness of plastics is such that they have become essential materials from domestic life to even space travel. The ease with which polymers can be processed, their cost-effectiveness, their high strength-to-weight ratio, their compatibility with biowastes to form composites and aesthetic values of finished products are some of the reasons why they are the most sought-after and used materials today. Polymers are used for various applications and some of the major areas are packaging, consumer products, transportation, medical, apparel, comfort applications and building construction, just to name a few.

The author with his hands-on experience spanning over 50 years in many diverse industries and also having pioneered some manufacturing industries after his experience with chemical giants such as BASF, ICI, Bayer AG, Hoechst AG and BP Chemicals Ciba Geigy makes this presentation both interesting and valuable, giving a reader a thorough in-depth knowledge of processing of polymers. The inclusion of a product manufacturing example for some important processes based on actual practice will no doubt be of great interest and useful to readers. Also included is a chapter on specialty polymers, their applications and current trends. This book has been designed to present the practical aspects of processing of polymers, rather than based on the theory of polymers.

The author has had the privilege of setting up manufacturing plants for products with a variety of polymers in countries such as Sri Lanka, Canada and the Philippines and also has helped many entrepreneurs to start from scratch. His expertise also includes foreign assignments for a Canadian agency to assist companies in Russia, Trinidad, Philippines and Serbia in the fields of polymer technology, manufacturing technology, increasing process efficiencies and troubleshooting and waste management. For the benefit of the readers, this presentation includes information from these experiences also, dealing with efficient processing of different types of important polymers.

The author thanks Lena Stoll, Acquisitions Editor, and Dr. Ria Fritz, her working colleague, at De Gruyter for their wonderful support and cooperation extended to him in compiling this presentation. The author hopes this book will be of immense

https://doi.org/10.1515/9783110656152-202

value and interest to students, teachers, consultants, researchers, manufacturers and also to entrepreneurs as it presents a thorough knowledge of the major aspects of processing polymers.

Chris Defonseka

# Contents

# Chapter 1
# Introduction to Polymers

## 1.1 The World of Polymers

The world of polymers is both exciting and challenging and with increasing possibilities due to constant research and development programs. Polymers were in use from ancient times in the form of gums, sealants, adhesives and so on, although their real value was realized only many years later. Over the years, plastics derived from polymers, mainly from crude oil as a starting sources, branched out into many different products as the demand and need for various items expanded in people's daily lives. Gradually at first and then rapidly, the world realized the advantages of using plastics and since then plastics have been replacing traditional materials. However, over the years, as air pollution grew mainly due to industrial activity, scientists and chemists have been searching for alternate sources for polymers and are discovering newer fields of nonpetro-based sources for polymers and additives to counter these environmental hazards.

A polymer is a large molecule (macromolecule) composed of repeating structural units, typically connected by covalent bonds. The basic unit of a polymer is a **mer** and *poly* means many, derived from the Greek word *polymeros*. So, a polymer can be defined as a substance with many basic units. These basic units are called **monomers.** There are different types of monomers and the process by which they are joined together is called **polymerization.** Some of the basic monomers are **ethylene monomer** and **styrene monomer** to form polyethylene (PE) and polystyrene (PS), respectively.

If the same types of mers are joined together, they are called "homopolymers" and if two different types of mers are joined together, they are called "copolymers." When three different types of mers are joined together, they are known as "terpolymers." Then again, these polymers are classified into two main groups: *thermoforming* and *thermosetting* polymers. Thermoforming polymers will soften on heating and after molding can be recycled and used again a number of times, depending on the degree of degradation of material and color issues. Thermosets, on the other hand, cannot be reused after molding. However, some of them can be recycled into a different product, for example, polyurethane foam wastes can be shredded and rebonded into a material suitable for carpet underlay, mattress bases and other applications. Some examples of thermoforming polymers are PE, PS and polypropylene, while examples of some thermosets are melamines, silicones and polyurethanes.

Most polymers are derived from petroleum-based ethylene gas as a biproduct from refining of crude oil. However, due to growing environmental hazards, the plastics industry has been carrying out intense research to move away from this source, and scientists and resin developers have been coming up with some hopeful

https://doi.org/10.1515/9783110656152-001

alternate sources but they will take some time before they can match the source from crude oil. A good example is: polyols from soybean oil, where they are used to make good quality polyurethane foams, although with lesser yield and carries an odor which has to be masked. These challenges are being addressed. The use of biowaste-filled polymer composite resins (40–60%) will also reduce the use of petro-based chemicals. Another good example is bamboo textiles – bedsheets, towels, pillow covers and so on – which can be mixed with small quantities of synthetic polymers such as rayon, polyester and also natural fibers from cotton and cellulose. Actual use and tests have shown that they are equally good, if not better than traditional textiles, with the additional advantage of less cost.

Because of their versatility and extraordinary range of properties, polymeric materials have established themselves as an essential part of everyday life, ranging from the very familiar plastics and elastomers to natural biopolymers such as nucleic acids and proteins that are essential for life. There are other varieties of natural polymers, for example, cellulose, which is the main constituent in wood and paper.

Polymers are generally available as solids, liquids, powders and other forms, either colored or in natural color, as small to medium packs, in steel drums or in bulk packages in paper bags or large totes. For large volume manufacturers, larger delivery systems are available. Additives are generally supplied as powders, liquids, color master batches and so on.

It is interesting to note that in addition to finding alternate sources for polymers, emerging technologies are establishing the use of nontraditional fillers and stiffening agents from natural sources such as bamboo fiber, rice hulls powder (lignin/silica), wheat husk flour (lignin/silica), egg shells powder (calcium carbonate) and graphene in manufacturing composite polymer resins and polymer composites. This makes the polymer industry less dependent on petroleum-based chemicals, thus helping to lessen environmental concerns. The recent discovery of the uses of graphene has opened out vast possibilities for the polymer industry with special electrically conductive polymers, high-strength-to-lightweight panels for the automotive industry, much more efficient and thinner cost-effective coatings for solar energy substrates than the traditional thicker cadmium coatings, just to mention a few practical applications.

One of the greatest developments in the polymer industry in recent years is the manufacture of polymeric composites with biocomposite wastes, particularly rice hull waste powder or flour. The earlier manufactures were called wood polymer composites (WPCs), where wood wastes were combined with polymers producing a material as a substitute for natural wood. However, PCRs (author's nomenclature) with rice hulls are emerging as even better materials than WPCs with applications on a wide spectrum including some applications – railway sleepers, flooring, kitchen cupboards, office furniture, building construction and many more – as an ideal substitute for natural wood for the building trades also.

Another major breakthrough for the plastics industry in recent years is the production of diesel oil from plastic wastes. There are now plants in some countries where these productions are taking place. Take the case of Huayin Energy Company in China. They use a breakdown process using pyrolysis technology to convert plastic wastes into liquid fuel and gases. In this high-tech **continuous pyrolysis** process, the plastic wastes produce around 50–75% fuel oil, 30–35% carbon black and 8–10% flammable gases. These projects are not only profitable businesses but more importantly help to reduce dependency on global supplies of diesel oil, reduce landfills and disposal problems, thus helping global pollution in an effective way.

Although in the west, plastic wastes are posing recycling challenges, probably because of the sheer volumes being generated, in Asia and the far-east, many new technological concepts are successfully using all grades of plastic wastes in such manufactures like floor tiles, roofing tiles, roofing sheets, building trades, polymer-modified bitumen in road paving, adhesives and many other applications. The additional use of nontraditional fillers and stiffening agents as mentioned earlier makes excellent quality products and cost-effective over traditional ones. For example, the use of 40% plastic wastes mixed with bitumen in road paving can absorb huge amounts of plastic wastes and it is found there is a tremendous improvement in surface quality and durability as well. Versatile bamboo fiber can easily replace the use of glass fiber or other fibers in the fabrication of fiber-glass products. Rice hulls ash can be used in concrete as a filler and as a strengthening agent. The author suggests that the inclusion of a small amount of rice hulls or wheat hulls powder as a flour or ash will make the road surfaces last longer because of the high presence of silica in the rice hulls acts as an excellent moisture barrier and gives additional strength. This will not affect the flexibility of the road surface created by the addition of plastic wastes but instead gives it additional flexibility.

With the rapid growth of the polymer industry, where newer polymers, especially specialty polymers are emerging, the need for more efficient and cost-effective processing machinery and equipment will no doubt keep the engineering designers and machinery manufacturers facing more and more exciting challenges. One area where specialized equipment will be needed will be to process composite polymer resins and polymeric composites that are finding great applications particularly in the automobile, aviation, transport, building construction industries and even for space travel. The polymer industry is well organized with advances taking place frequently in many directions. Manufacturers of polymer resins and processors adhere strictly to accepted international standards and with many industrial associations backing for maintaining standards and quality.

When enumerating the value and importance of the vast possibilities with polymers, one must also give thought to methods of disposing of wastes with traditional methods of landfills and incinerating (which releases toxic gases) as it is not viable in our present-day world, where the demand for plastics is increasing one way or the other. It is heartening to note that viable solutions are available with newer concepts of disposal, recycling and reusing methodologies such as polymer-modified bitumen, where huge quantities of plastic wastes can be directly mixed with hot bitumen in road paving, resulting in better road surfaces, polymeric composite lumber, where plastic wastes/polymers and biowastes like rice hulls, wheat hulls and others can be combined to produce materials as an ideal substitute for natural wood and also production of diesel oil from plastic wastes using a special method called **continuous pyrolysis** with carbon as the residue, which can be used for other manufactures. Considering these large-scale usages of plastic wastes and coupled with the traditional recycling of plastic wastes into reusable pellets, plastic wastes could be well controlled.

## Bibliography

[1]    Defonseka, Chris. "Practical Guide to Water-Blown Cellular Polymers," P. 1, Smithers Rapra, 2016.

# Chapter 2
# Basic Chemistry of Polymers

## 2.1 What Are Polymers?

Polymer chemistry is a complex subject and the basic information presented here is sufficient for the readers to gain a good understanding of polymers, which are the backbone of the plastics industry.

Chemistry is the science that deals with the composition of matter and how changes take place. The two broad areas of chemistry are *organic* and *inorganic*. Organic chemistry deals with matter that contains the element carbon, while the latter deals with matter which is mineral in origin. The term organic was originally used to mean compounds of plant or animal origin but now it also includes many synthetic materials that have been developed through constant research and development. One such group of synthetic materials is called *plastics*.

The word plastics is derived from the Greek word – *Plastikos* – and a plastic material can be generally defined as a material that is pliable and capable of being shaped by temperature and pressure. Plastics are based on polymers, sometimes called *resins*, which is not correct, since a resin is a gum-like substance.

Polymers are materials composed of molecules of very high molecular weights. These are generally referred to as *macromolecules* which have structures that are generated synthetically or through natural processes. Polymers are formed when basic monomer units called *mers* are joined together during processes called *addition* or *chain-growth polymerization* and *radical* or *condensation polymerization*. In addition polymerization, the final molecule is a repeating sequence of blocks with a chemical formula of the monomers. Condensation polymerization processes occur when the resulting polymers have fewer atoms than those present in the monomers from which they are generated. Since many additional polymerization processes result in condensates and various condensation polymerization processes are chain-growth polymerization processes that resemble addition polymerization, polymerization processes, they may be also broken down into *step polymerization* and *chain polymerization*.

Linear and nonlinear step-growth polymerization are processes where the polymerization occurs with more than one molecular species. On the other hand, chain-growth polymerization process occurs with monomers with a reactive group at the end. Free-radical polymerization is the most popular polymerization method used and it is used to polymerize monomers with general structures.

There are various ways in which monomers can be arranged during polymerization and they can be broken down into two general categories: noncross-linked and cross-linked. Furthermore, the noncross-linked can be subdivided into linear and branched polymers. The most common example of a noncross-linked polymer that

https://doi.org/10.1515/9783110656152-002

presents the various degrees of branching is polyethylene (PE). Another important group of this same category is copolymers, which have two or more different mono-mer types in the same chain. Depending on how the different monomers are arranged in the polymer chain, they are called *random, alternating, block* or *graft* copolymers.

## 2.2 Polymer Categories

Polymers are generally classified into six categories based on source, light penetra-tion, heat reaction, polymerization reaction and crystal structure.

Source: Polymers from natural sources, modified natural or synthetic polymers.

Light penetration: Optical properties, for example, opaque, transparent, trans-lucent or luminescent

Heat reaction: Thermoplastic and thermosetting

Polymerization: Method of joining basic monomers to form polymers, copoly-mers and terpolymers

Crystal structure: Crystalline (molecules arranged in order) and amorphous (random arrangement of molecules)

To understand different materials, their properties and general processing methods for polymers, one must have a basic understanding of the chemical structure of plastics. According to general chemical theory, atoms which forms molecules, con-sists of neutrons, protons and electrons. The atoms of different elements contain different numbers of these subatomic particles and, thus, no two elements have identical atoms.

Neutrons with no electrical charge and protons that are positively charged form the nucleus of an atom, while the electrons that are negatively charged are located in orbits or shells around the nucleus. For all elements, the atom itself is electrically neutral. If an atom gains an electron, it becomes *electronegative* (a negative ion), and if it loses an electron, it becomes *electropositive* (a positive ion). An ion is an atom that has either gained or lost one or more electrons.

All elements (e.g., carbon, oxygen, hydrogen) are classified in the periodic table according to their atomic number, which is the number of protons in the nu-cleus of an atom. The atomic weight of an element is the weight (mass) of an atom of that element compared with the weight mass of an atom of carbon. Some ele-ments are very active chemically, while some others are inert and will not readily combine chemically. In the study of polymers and plastics some of the elements that are of primary importance are carbon (C), oxygen (O), hydrogen (H), chlorine (Cl), fluorine (F) and nitrogen (N). These form the backbone of converting polymers into different plastic materials.

Most elements are combinations of two or more atoms bonded together to form molecules. In plastics, the bonding patterns are very important since these determine the physical characteristics of the final plastic product. Bonds can be divided into primary bonds (ionic or covalent bonds) and secondary bonds (van der Waals forces). All bonds between atoms and between elements are electrical in nature and most elements constantly try to reach a stable state by (a) receiving extra electrons, (b) releasing electrons or (c) sharing electrons.

As we know, a polymer is composed of several basic units called *mers* which are considered as bifunctional or difunctional with two reactive bonding sites. Monomers are the basic units that form different polymers and, in most cases, are liquids.

One of the most important monomers in the plastics industry is ethylene, which is a gas as a bi product of refining crude oil. Liquefied ethylene gas is the starting source for many plastics. The terms plastics and resins are used interchangeably but such use is actually incorrect. A resin in a plastic determines whether that plastic material is thermoplastic or thermosetting. A resin does not become plastic until polymerization occurs and the product is in the finished state. The classification – plastics – does not include a large category of elastomers such as synthetic rubbers. However, the term polymer includes both plastics and elastomers. An elastomer can be defined as a substance that can stretch more than 200% and return back almost to its original size.

Polymers can be radically altered by copolymerization, cross-linking, orientation and processing. The incorporation of additives such as fillers, heat stabilizers, stiffening agents, colorants and lubricants will produce materials for general-purpose applications, while additional special additives will produce what is known as "engineered polymers." These special polymers are for specific special applications, where particular physical and chemical properties are required.

## 2.2.1 Polymer Microstructures

The microstructure of a polymer (sometimes called a configuration) relates to the physical arrangement of monomers along the backbone of the polymer chain. To change, these are the elements of a polymer structure where the covalent bonds have to be broken. A structure has a strong influence on the properties of a polymer.

An important microstructural feature determining polymer properties is the "architecture" of the polymer. A simple polymer architecture is a linear chain; a single backbone with no branches. A branched polymer molecule is composed of a main branch point with one or more chains or branches. Branching of polymers chains affects the ability of chains to "slide" past one another by altering intermolecular forces, which in turn affects the physical properties of a polymer. Long-branch chains may increase polymer strength, toughness and glass-transition temperature ($T_g$) due to an increase in the number of entanglements per chain. Conversely, random lengths and short chains may reduce polymer strength and also reduce crystallinity of a polymer. A good example of this effect is related to the physical attributes of PE.

High-density PE (HDPE) has a very low degree of branching, is quite stiff and is used to make products such as containers, jugs and boxes. Low-density PE (LDPE) has a significant number of long and short branches, is quite flexible and is used in applications such as films.

### 2.2.2 Lengths of Polymer Chains

Physical properties of a polymer are strongly dependent on the size or length of the polymer chain. For example, as chain length increases, melting point and boiling points increase, as does the impact strength and viscosity (resistance to flow) of a polymer in its melt state. A tenfold increase in the length of a polymer chain results in a viscosity increase by > 1,000 fold. Increasing chain lengths tends to decrease chain mobility, increase strength and toughness and increase the $T_g$. A common means of expressing the chain length of a polymer is the "degree of polymerization" which quantifies the number of monomers incorporated into the chain. As with other molecules, the size of a polymer may also be expressed in terms of molecular weight.

### 2.2.3 Polymer Morphology

Polymer morphology can be described as the arrangement of chains and the order of the many polymer chains. A synthetic polymer may be described as "crystalline" if it contains regions of three-dimensional ordering on scales of atomic (rather than macromolecular) length. Synthetic polymers may consist of crystalline and amorphous regions, and the degree of crystallinity is expressed in terms of a weight fraction or volume fraction of crystalline material.

Polymers with microcrystalline regions are, in general, tougher, can be bent more without breaking and more impact-resistant than totally amorphous polymers. Polymers with a degree of crystallinity approaching zero or one tend to be transparent, whereas polymers with intermediate degrees of crystallinity tend to be opaque due to light scattering by crystalline or glassy regions. Thus, for many polymers, reduced crystallinity may also be associated with increased transparency.

### 2.2.4 Polymer Behavior – Melting Point

The melting point of a material would generally suggest a solid-to-liquid transition. However, if applied to polymers, the melting point is more a transition from a crystalline phase to a solid-amorphous phase. This behavior is more appropriately called the "crystalline melting temperature." Among synthetic polymers, crystalline melting is discussed only with respect to thermoplastics, because thermosetting polymers decompose at high temperatures rather than melt. The boiling point of a polymer is strongly dependent on chain length. Polymers with a large degree of polymerization do no exhibit a boiling point because they decompose before reaching theoretical boiling temperatures. For polymers with shorter chain lengths, a boiling transition may be observed.

### 2.2.5 Polymer Behavior – Mixing

Polymeric materials are far less miscible as mixtures than materials with small molecules. The driving force for mixing is usually entropy (disorder or randomness of a system) not interaction energy, meaning that miscible materials usually form a solution because of an increase in entropy, and hence, free energy is associated with increasing the amount of volume available to each component. Polymeric molecules are much larger and have much larger specific volumes than smaller molecules, so the number of molecules involved in a polymeric mixture is far smaller than the number in a small molecule mixture of equal volume. Furthermore, the phase behavior of polymer solutions and mixtures is more complex than that of small-molecule mixtures.

Incorporation of plasticizers tends to lower the $T_g$ and increase polymer flexibility. Generally, plasticizers are small molecules that are chemically similar to polymers and create gaps between polymer chains for greater mobility and reduced interchain interaction. A good example of plasticizer actions is polyvinyl chloride (PVC). Unplasticized PVC (uPVC) is used for the production of pipes, which has no plasticizer in them as they have to be strong and heat resistant. If plasticized PVC is used for pipes, over a period, it will lose the plasticizer due to evaporation due to heat and will become brittle and split. Plasticized PVC is generally used for flexible applications such as hose, films, sheeting, coatings and so on.

## 2.3 Polymer Blends

Polymer blends are polymeric materials which are made by mixing or blending two or more polymers to enhance the physical properties of each individual component. Some common blends are polypropylene–polycarbonate (PP–PC), polyvinyl chloride–acrylonitrile butadiene styrene (PVC-ABS), PC–polybutylene terephthalate (PBT) and PE–polytetrafluoroethylene (PTFE). The process of polymer blending is done by dispersion of a minor or a secondary polymer within a major polymer that serves as the matrix for the whole blend.

Blending can also be called as mixing, which is subdivided into distributive and dispersive mixing. The morphology development of polymer blends is generally determined by three mechanisms: distributive mixing, dispersive mixing and coalescence.

## 2.4 Polymer Composites

Composites can be defined as materials made from two or more constituent components, with significantly different properties, that when combined produces a final material with characteristics different from the individual components, while remaining separate and distinct within the structure. Generally, one main component will act as the matrix. These versatile composite materials can be formulated to give superior properties and higher structural strengths. These properties can also be optimized by the addition of suitable additives. Due to global environmental concerns and rising polymer prices, researchers and developers have been coming out with composite polymer resins.

Some of the common polymers used are LDPE, HDPE, PP and PVC. These materials can be termed as: *thermoplastic bio-composites*. This subject will be discussed in a separate section later on.

## 2.5 Cellular Polymers

Cellular polymers, also known as "foamed polymers" or "expanded plastics," are very important to human life. Cellulose is the most abundant of all naturally occurring organic compounds with wood as a cellular form of the polymer cellulose. This name is derived from the Latin word *cellula,* meaning very small cells or very small rooms. Most polymers can be foamed to give cellular structures but only a few have found favor with commercial applications.

Some of the common polymers used for foaming are PE for packaging, polyurethane (PU) for comfort, polystyrene (PS) for packaging and insulation and PVC for floating devices, sheeting, artificial leather and electric cables. In most cases, a blowing agent is incorporated during polymerization which will later expand the polymer under heat and pressure. Polymer expansions can be done with either physical or chemical blowing agents. As an example, inert pentane gas is used for expandable PS which will turn the polymer into expanded foam under heat but new trends are the use of water as a blowing agent for cellular polymers.

The need for cellular foams are ever widening with the expansion of air travel, automobile industry, building construction, packaging, consumer items, bedding, furniture, fishing industries and many others, including space travel where special viscoelastic foams are used for comfort to counter "G" forces.

## 2.6 Engineered Plastics

Engineered plastics encompass the engineering of polymeric materials to suit high-end special applications. Basics are designing and development of properties such as mechanical, chemical, physical and processing ability of the engineered materials to manufacture products suitable for applications they were designed for. Also important are factors such as weatherability, thermal properties, electrical properties, durability, barrier properties and resistance to chemical attack. In some cases, aesthetic values also may be important.

In plastics engineering, as in most engineering disciplines, the economics of a final product plays an important role and beginning with the cost per lb/kg of material and processing costs are also significant factors. An engineered material may meet all rigid required specifications but if the processing or the processed products are cost-prohibitive, the material may be considered as not viable.

A big challenge for plastic engineers is final disposal and reduction of the used materials. Some of the areas where engineered plastics are required are:
- Consumer needs
- Medical needs
- Automotive parts
- Space travel
- Building construction
- Biodegradable plastics
- Elastomers/rubbers
- Epoxies
- Transport

Some of the processes used for manufacturing products are: injection molding, extrusion, compression molding, stretch blow molding, laminating, fiberglass molding, filament winding, vacuum forming and rotational molding.

## 2.7 Bioplastics

Bioplastics are plastics derived from renewable biomass sources such as vegetable oils and fats, corn starch, straw, woodchips and others. Bioplastics can also be made from agricultural by-products and also from used plastics bottles and other containers using microorganisms. Bioplastics are usually derived from sugar derivatives, including starch, cellulose and lactic acid.

Bioplastics are used for low-end disposable items such as packaging, cutlery, crockery, pots, bowls and straws. Few commercial applications are viable due to costs and poor performance to meet high-end applications specifications. However, ongoing research and development programs are hopeful for better materials in the future.

In pursuit of mitigating climate change and promoting "green-technology," a company called Bio Carbon 3D Ltd. in British Columbia has developed a process to make high-performance bioplastics from wood. Rather than another bioalternative for products like straws and containers, this is wood-based engineered grade for high-performance plastics. People often think of bioplastics as single-use materials with low functionality but these new materials being produced have high functionality with exceptional properties like incredibly high-heat resistance and being lightweight.

The process in brief is the extraction of the polymer resin from the wood by mixing with a solvent and put through a series of pressurized-heating and cooling phases till the polymer is extracted. Initially, the company plans to produce a number of different filaments and then additional grades with carbon fiber reinforcements and conductive filaments. As the technology and further developments takes place, these bioplastic materials will have a great potential for many high-performance end applications.

## 2.8 Important Common Polymers

Although the polymer spectrum is vast, there are a few important polymers which produce essential products for daily life of people. They are PE, PS, PP, PVC, PC, PU, melamines and others. The first five are thermoforming polymers, while the last two are thermosets. These will be discussed in detail in Chapter 3.

## 2.9 Polymer Degradation

Polymer degradation refers to a change in properties such as color, tensile strength, shape and molecular weight. This can be caused by heat, light, chemicals, extreme cold, long shelf lives, and in some cases, galvanic action. Very often polymer degradation occurs due to the separation of the bonds due to the action of water (hydrolysis) connecting the polymer chains, which results in a decrease in molecular mass of the polymer. During processing of polymers, degradation can take place due to excessive heat or dwell times. In some cases, where dyes, pigments or master batches are used for coloring, degradation can take place due to nonuniform dispersion.

Carbon-based polymers are more prone to thermal degradation than inorganic polymers, and therefore, may not be suitable for most high-temperature processing. While thermosets cannot be reused after heat processing, even thermoforming polymers although reusable has limits to how many times it can be recycled due to gradual degradation.

## 2.10  Polymer Waste Disposal

As the world population grows, the demand for "plastics" is growing by leaps and bounds and also due to alternate sources for crude-oil has been emerging for some time now, the plastics industry has gained more strength. However, it has to face growing environmental concerns as the traditional disposal of wastes through land-fills and burning is no longer viable due to the sheer volumes involved. This subject will be discussed in detail later under "Recycling."

## Bibliography

[1]   Osswald, Tim A., Baur, Erwin, Brinkmann, Sigrid, Oberbach, Karl, Schmachtenberg, Ernst. "International Plastics Handbook", P. 17, 53, Hanser Publication, 2006.
[2]   Atik, Tamar. "Developing high grade performance bioplastics from wood", P. 16, Canadian Biomass Magazine winter, 2019.
[3]   Defonseka, Chris. "Practical Guide to Cellular Polymers", P. 1–5, Smithers Rapra, 2016.

# Chapter 3
# Types of Polymers

## 3.1 Common Polymers

There are many polymers that are converted to final plastic materials by processing but a few like polyethylene (PE), polystyrene (PS), polypropylene (PP), polycarbonate (PC), polyvinyl chloride (PVC), melamine, polyurethanes and epoxies stand out as the most used polymers to meet essential daily needs of people. The discovery of graphene has opened a whole lot of exciting possibilities for the plastics industry. Its strength to lightweight properties, high thermal conductivity and others bodes well for the automobile industry, in particular. Emerging technologies like water-blown cellular polymers and biocomposite fillers for polymers like bamboo fibers, rice hulls, wheat hulls and coconut shell powders to name a few also provide wider scope of applications for polymers. Two important factors stand out here: less use of petro-based products and cost-effectiveness.

### 3.1.1 Polyethylenes

PE is a thermoplastic polymer with variable crystalline structure and an extremely large range of applications. It is one of the most widely produced polymers in the world and is commonly categorized into one of several major compounds. The most common PE includes low-density polyethylene (LDPE), linear low-density polyethylene (LLDPE), high-density polyethylene (HDPE) and ultra-high-molecular-weight polyethylene (UHMWPE). Other variants include medium-density PE, ultra-low-molecular-weight PE or PE Wax, high-molecular-weight PE, high-density cross-linked PE, cross-linked PE, very low-density PE and chlorinated PE.

- LDPE – It is a very flexible material with unique flow properties that makes it particularly suitable for plastic film applications like shopping bags and packaging film. LDPE has high ductile but low tensile strength which is evident by its propensity to stretch when strained.
- LLDPE – Very similar to LDPE with the added advantage that properties of LLDPE can be altered by adjusting the formula constituents and that the overall production process is less energy intensive than LDPE.
- HDPE – It is a strong, high-density, moderately stiff plastic with a high crystalline structure. It is frequently used as an ideal material for milk, cartons, laundry detergent packs, garbage bins and many other applications. HDPE is a popular matrix for composite polymer resins where about 40% can be filled with biocomposites.

https://doi.org/10.1515/9783110656152-003

- UHMWPE – It is an extremely dense version of PE with molecular weights typically in an order of magnitude greater than HDPE. It can be spun into threads with tensile strengths greater than steel and is frequently incorporated into materials for high-performance equipment like bullet-proof vests.

PE can also be "blown" to produce cellular PE foams which are popular in packaging and padding. For industrial use these foams are made in different colors with black being probably the most popular.

### 3.1.2 Polystyrenes

A PS is a versatile, hard, stiff and transparent synthetic polymer resin produced by polymerization of styrene monomer. The presence of pendant phenyl groups is key to the properties of PS. Solid PS is transparent owing to these large ring-shaped molecular groups, which prevent polymer chains from packing into close crystalline prebonds, making the polymer rigid and hard. PS can also be blended or copolymerized with other polymers to obtain high-impact strength or even blended with rubber for more flexibility.

Probably, the most versatile and widely used insulation and packaging material in the world is cellular PS – *expandable polystyrene* (EPS) – which is a closed-cell foamed material. EPS was first made with chlorofluorocarbons as blowing agents but has been banned for some time now due to environmental concerns. Current practices use pentane gas incorporated into PS during polymerization, resulting in very tiny beads with each containing small amount of the blowing agent. The bead size can be varied to suit customer preference and end applications. If colored beads are required, the coloring material must be incorporated at the time of polymerization to obtain maximum color effects. It is also possible to color the beads during preexpansion but only surface coloring with a "mottled" effect will result.

More recent developments have made it possible to produce equally good cellular PS foams as EPS with water as the blowing agent with emission of harmless water vapor instead of harmful gases. This will also reduce the fire-hazard factor due to static and friction during the shelf life of the beads. This material is known as *WEPS* and the only difference in processing is at the preexpansion stage, and slightly different preexpanding conditions are required.

Another grade of PS used in very large volumes for insulation and in the building trades is *XPS* – an extruded grade where a standard grade of PS is extruded and incorporated with a gas as a blowing agent during extrusion. This process will produce cellular boards around 1 m in width and cut to desired lengths on the takeoff conveyor.

While standard grades of PS are used for making consumer products, EPS and WEPS cellular foams will be used for packaging, insulation, sound-proofing, fishing industry, building construction and others, and modified grades with synthetic rubber, for example, acrylonitrile butadiene styrene (ABS) are used for pipes, computer housings, refrigerator housings and many other applications.

### 3.1.3 Polypropylenes

PP is a synthetic polymer resin made by polymerization of propylene. As an important polymer of the polyolefin family, it can be processed in many ways. Propylene is a gaseous compound resulting from the thermal cracking of ethane, propane, butane or naphtha originating from petroleum fractions. Like ethylene it belongs to a "lower olefin class" of hydrocarbons, whose molecules contain a single pair of carbon atoms linked by a double bond. Under the action of the polymerization catalysts, however, the double bonds can be broken and thousands of propylene molecules will link together to form chain-like polymers. Propylene can be also be polymerized with ethylene to produce elastic ethylene–propylene copolymer.

One may consider PP as a "higher grade" polymer of PE due to some of its special properties. Its molecular structure is such that it can stand repeated flexing without cracking. This special property of fatigue resistance coupled with its other versatile properties makes it a very valuable polymer. PP can be blow-molded, extruded, injection molded and also made into film. A large proportion of PP polymers are made into yarn for textiles, ropes and films for packaging.

Some of the other products made from PP are used for housing appliances, dishwasher-safe food containers, toys, automobile battery cases, outdoor furniture, cordage, protective wear, medical applications, fibers for playing fields as a substitute for natural grass, agricultural applications and building construction work. Due to its excellent properties such as toughness, resilience, water resistance and chemical inertness, PP have a wide spectrum of industrial and engineering applications also.

Expanded (or cellular) PP are also very popular and have an expanding global market. These foamed materials are used widely in packaging, bedding, furniture, carpet underlays and so on.

### 3.1.4 Polycarbonates

PC plastics are a naturally transparent amorphous thermoplastic material with a light transparency factor similar to glass. Although these materials are commercially available in a variety of colors, the most popular grades are the fully transparent ones. PC polymers are used for making a variety of products and are particularly useful when impact resistance and/or transparency are important in a product – for example, bullet-proof glass.

PC being an amorphous material does not exhibit the ordered characteristics of crystalline solids. Typically, amorphous plastic materials demonstrate a tendency to gradually soften, meaning they have a wider range between their glass transition temperature and their melting point, rather than to exhibit a sharp transition from solid to liquid as is the case of crystalline polymers. PCs are copolymers in that they are composed of several different monomer types in combination with one another.

PCs are incredibly useful plastics for applications requiring transparency and high-impact resistance. It is a lighter alternative to glass and a natural UV filter. In the modern current era, PC are produced by a large number of companies, typically with their own production processes and unique formulae. Trade names include well-known variants for PC resins such as Lexan by SABIC and Makrolon by Bayer AG. There are various industry grades of PC available, with most called by the generic name (PC) and are typically differentiated by the amount of glass-fiber reinforcing they contain and the variance in melt flow between them. Some may have additives such as "ultraviolet (UV) stabilizers" to protect the material from long-term exposure to UV light. Molding grades may have other additives such as mold release agents and lubricants to enhance processing. PC materials have very good heat resistance and can be combined with flame retardant materials without significant material degradation.

Comparison of relative impact strengths of plastics materials like PVC, PS, acrylics, and ABS with PC material will show that PCs will have much higher values than them. Another feature of PC is that they are very pliable. It can typically form at room temperature without cracking or breaking, similar to aluminum metal. Although deformation may be simpler with the application of heat, even small angle bends are possible without it. This characteristic makes PC sheet stock particularly useful in photocopying applications, when transparency is required or when nonconductive material with good electrical insulation properties is required.

PC material in sheet or round stock form can be machined on a mill or a lathe. Colors are usually limited to clear, white and black and in some cases tinted or translucent colors. Parts that are machined from clear stock usually require some postprocessing finishing to remove tool marks and to restore the transparent nature of the material. Since PC is a thermoplastic material, certain three-dimensional (3D) printers are able to print with PC in filament form with the 3D printer heating and depositing the filament into desired shape. PC/ABS blends are also available for 3D printing and are usually limited to white color.

PC is commonly used for bottles, plastic lenses (with scratch-resistant coatings), in medical devices, automotive components, protective gear, greenhouses, digital disks, exterior e.

### 3.1.5 Polyvinyl Chloride

PVC is a very versatile and widely used polymer in domestic, industrial, automotive, building construction, packaging and agricultural sectors. PVC is a polymer resin made from polymerization of vinyl chloride (VC) and is a much needed polymer resin as the other popular ones. VC is also known as "chloroethylene" and is a lightweight rigid plastic in its pure form and is also available in flexible grades in its "plasticized" form. VC is usually obtained by reacting ethylene with oxygen and hydrogen chloride using a copper catalyst.

PVC is obtained by subjecting VC to highly reactive compounds known as "free-radical initiators." Under the action of these initiators, the double bonds in the VC monomers are opened and one of the resultant single bonds is used to link together thousands of VC monomers to form the repeating units of the polymers, which are large multiple-unit molecules. Unplasticized PVC which is rigid are used for rigid applications and plasticized PVC are used for flexible applications.

Pure PVC is used in applications in the construction trades, where its rigidity, strength and flame-retardant properties are useful for pipes, conduits, window sidings and frames and also door frames. Rigidity can be altered by the addition of plasticizers and highly flexible PVC can be processed in many ways such as extrusion, blown film sheets, cast into floating devices, used as insulation coating for electrical cables, coatings for artificial leather, dip coating, sheets for protective wear and many other applications. PVC is also one of the suitable polymers for using as the matrix for composite resins and wood polymer composites or polymeric composites with rice hulls.

### 3.1.6 Polyurethanes

Polyurethanes can be basically categorized as flexible, semirigid, rigid, microcellular, viscoelastic and thermoplastic urethanes. Of these, the flexible grades are the most popular materials. Polyurethane materials as PU foams essentially differ from most other plastics in that there is no urethane monomer as such and this is invariably formed during the manufacture of a PU foam product. In general, they are made by reacting a polyol with an isocyanate and other additives such as blowing agents, stabilizers and fillers. Polyols can be either based on polyether or polyester. If very soft foams are desired, a secondary blowing agent like methylene chloride can be used. Since this process is "exothermic" (heat-giving), extra precautions are needed during manufacture and also postcuring for at least 24 h.

Polyols give soft open-cell foams, and with the incorporation of suitable additives, special properties as required for end applications can be achieved. Some of the common special properties required by the consumer and industrial markets are UV protection, high resiliency, fire retardancy, antifungal, density and support factor. Polyether polyols will give soft foams, while polyester-based polyols will give less flexible and rigid foams due to their smaller cell structures.

A very special grade that belongs to the polyurethane foams family is *viscoelastic foams* which are four-dimensional (4D) and was originally created for space travel for cushioning to absorb and counter *g-forces*. While standard flexible foams used for comfort applications are two-dimensional, viscoelastic foams are 4D with properties such as *density, hardness, temperature and time.* This material subsequently became very popular especially among the consumer market for bedding and other applications due to its special properties that gives ultra-comfort, so much so, that it came to be called a *miracle foam.*

Since polyurethanes are made via chemical reactions with mostly petroleum-based components due to environmental concerns, scientists have been researching alternate sources to move away from petro-based sources moving toward natural oils and other alternatives. Natural oil polyols also known as biopolyols are derived from vegetable oils from soya beans, rapeseed (canola), jatropha plants and castor oils. Initially, manufacturers of polyols used them in combination with petro-based polyols but as technology advanced, they are now able to produce polyols from 100% oil-based sources with good results. However, still two factors have to be overcome – lesser yield and resulting odor, which has to be masked.

Polyurethanes are a very versatile and essential material in most spheres of human activity, including space travel, and being a thermoset cannot be reused but can be recycled and *compacted using either a steaming or adhesive processing methods, where the foam wastes are* shredded and made into large blocks for cutting sheets for carpet underlay, slabs for mattress bases, packaging and so on. Wastes can also be used to produce adhesives, varnishes, wood preservatives and others.

### 3.1.7 Melamine Formaldehyde

Melamine formaldehyde (MF) is a hard, very durable, heat resistant and versatile polymer resin belonging to the thermosetting group. It is made by condensation of the two monomers. MF resins are quite similar to urea-formaldehyde (UF) and are fully compatible with them and a resin blend is called melamine-urea-formaldehyde . Compared to natural veneers, particle boards made from these melamine composite resins display improved heat, moisture, scratch and chemical resistance. Melamine can also be converted to foam structures which are hard and used for insulating or soundproofing and also as abrasive cleaners. Unlike UF which is used mostly for molding electrical parts, MFs are widely used for dinner ware, cups and saucers and the like. Plain or printed MF dinner plates are very popular being almost unbreakable, high heat-resistant and for aesthetic values of finished products.

Melamine resins are generally used for the manufacture of many products including kitchenware, laminates, floor tiles, foamed cleaners/sponges and others.

Its versatility finds uses in insulation, soundproofing and also as a material incorporated as a fire-retardant additive. This factor includes manufacture of upholstery, thermal liners, heat-resistant gloves and industrial aprons and safety wear.

One of the most important applications of melamine is the fabrication of laminates and extrusion of particle board. To improve scratch resistant and fire retardant properties it can be blended with UF resins. Sulfonated melamine formaldehyde is used in industry as a super plasticizer in the manufacture of high-resistant concrete by the reduction of water but increasing fluidity.

Some of the major manufacturers of MF resins are Hexion, Arclin, Georgia-Pacific, Ineos, BASF and DIC under commercial names such as Saduren, Maprenal, Resimene and Leaf.

### 3.1.8 Acrylonitrile Butadiene Styrene

ABS is a terpolymer or a polymer consisting of three different polymers. This amorphous blend is made up of acrylonitrile, butadiene and styrene in varying proportions. Each of these monomers serve to impart a special property to ABS. Acrylonitrile provides chemical and thermal stability, butadiene increases toughness and impact strength and styrene gives the plastic material a nice and glossy finish.

Variations in the relative proportions of each monomer can result in drastic changes in the physical properties of ABS. As a general practice, the proportions of individual monomers may vary from 15% to 35% acrylonitrile, 40% to 60% styrene and 5% to 30% butadiene, resulting in a variety of ABS materials for plastic end products. It is also possible to make blends with others such as PVC, PC and polysulfonates.

ABS is a common thermoplastic and often is able to meet a wide range of requirements of many end applications at a reasonable price falling between standard PVC, PS resins and engineering resins acrylic and nylon. The primary advantage of ABS as a material is that it combines the strength and rigidity of acrylonitrile and styrene polymers with the toughness of the alkaline or acidic aging.

ABS is an amorphous solid, which means it technically has no melting temperature. For the process of melting, an approximate glass transition temperature is around 105°C. At this temperature, individual polymer chains take on a higher degree of conformation and can move and slide past each other. This property makes ABS a good material for injection molding. The temperatures at which molding is done also effects the properties of a product. Low temperature molding imparts greater strength and impact resistant, while molding at high temperatures makes glossier products with higher heat resistance.

ABS plastics became widely available in the 1950s and the variability of its copolymers and ease of processing made ABS one of the most popular engineering polymers. As a "bridge" polymer whose properties lie between commodity plastics and high-performance engineering thermoplastics, ABS has become one of the most used engineering thermoplastics.

## Bibliography

[1]  Flynt, Joseph. Article "Acrylonitrile Butadiene Styrene (ABS): A Tough and Diverse Plastic". https://3dinsider.com/whatisabs, Nov. 10 2017.

[2]  Defonseka, Chris. "Practical Guide to Water-Blown Cellular Polymers", P. 7–12, Smithers Rapra UK, 2016.

[3]  Polymer Properties Database: "Melamine-Formaldehyde Resins", website https://polymerdatabase.com, 2015.

[4]  Rogers, Tony. "Everything You Need to Know About Polycarbonate (PC)"-Creative Mechanisms, www.creativemechanisms.com, August 2015.

# Chapter 4
# Properties of Polymers

## 4.1 Importance of Properties

The properties of a polymer play a key role in determining its ability to achieve preset goals of a finished product. Although special properties can be achieved for polymers by the incorporation of additives, the attractive forces between polymer chains play a large part in determining the initial properties of a polymer. Generally, polymer chains are long, so these interchain forces are amplified beyond the attractions between conventional molecules. Different side groups can lead a polymer towards ionic bonding or hydrogen bonding between its own chains. These stronger forces, typically, will result, for example, in higher tensile strength and higher crystalline melting points.

Since the range of properties for polymers is wide, the author will present only selected important properties in relation to polymer processing. Some of the key properties of plastics are *specific gravity, density, water absorption, melt flow rate, mold shrinkage, tensile strength, flexural strength, impact strength, coefficient of thermal expansion, thermal conductivity and so on.*

## 4.2 Thermal Properties

When processing polymers, heat plays a major role and the heat flow through a polymer is controlled by conduction, determined by the thermal conductivity ($k$). However, at the onset of heating, the polymer responds as a whole until it reaches a steady-state condition and is generally controlled by its density and specific heat.

### 4.2.1 Thermal Diffusivity

This is the material property that governs the process of thermal diffusion over time. The thermal diffusivity factor in amorphous thermoplastics will decrease with temperature. A decrease in thermal diffusivity with increasing temperature is possible in semicrystalline thermoplastics.

### 4.2.2 Linear Coefficient of Thermal Expansion

The linear coefficient of thermal expansion is related to volume changes that occur in a polymer due to temperature variations. For many materials, thermal expansion is related to the melting temperature of that material. Similarly, there is also a

https://doi.org/10.1515/9783110656152-004

relation between the thermal expansion coefficient of polymers and their elastic modulus, while molecular orientation also will affect the thermal expansion of plastics.

### 4.2.3 Thermal Degradation

Since plastics are basically organic materials, they are threatened by chain breaking, splitting off of substituents and oxidation. A polymer material especially prone to thermal degradation is polyvinyl chloride (PVC). During this phase, the hydrogen chloride that results will attack metal parts and if inhaled can lead to health hazards. During processing, thermal degradation can be a critical factor. Direct contact with flames is considered extreme thermal loading of polymers. The flammability factor of polymeric materials is important in many common products such as clothing, construction materials and protective wear, to name a few.

## 4.3 Mechanical Properties

The mechanical or bulk properties of a polymer are of primary interest because they dictate how a polymer when converted into a plastic material will behave in actual use. Presented below are some of the important ones.

### 4.3.1 Tensile Strength

The tensile strength of a material will indicate how much stress the material will endure before undergoing permanent deformation. This property is very important in applications that rely on physical strength or durability of a plastic material. In general, tensile strength of a plastic material increases with the length of a polymer chain cross-linking with polymer chains.

### 4.3.2 Melting Point

The term "melting point" when applied to polymers suggests not a solid–liquid phase transition but a transition from a crystalline or semicrystalline phase to a solid amorphous phase. Among synthetic polymers, crystalline melting is discussed only in regard to thermoplastics because thermosetting polymers decompose at high temperatures rather than melt.

### 4.3.3 Young's Modulus of Elasticity

Young's modulus quantifies the elasticity of a polymer. It is defined as the ratio of the rate of change of stress to strain. In polymer applications, this is a highly relevant property when the physical properties of a polymer are important. The modulus is strongly dependent on temperature. Dynamic mechanical analyzes are used to measure the complex modulus by oscillating a load and measuring the resulting strain as a function of time.

### 4.3.4 Glass Transition Temperature

In the manufacture of synthetic polymers, an important aspect is the glass transition temperature or point ($T_g$). This is the temperature at which amorphous polymers undergo a transition from a rubbery, viscous liquid to a brittle, glassy amorphous solid. The $T_g$ can be engineered by altering the degree of branching or cross-linking in a polymer or by the addition of plasticizers, which lowers the $T_g$ and increases polymer flexibility. A good example of the influence of plasticizers is the case of PVC. Unplasticized PVC is used for manufacturing pipe because it needs to be strong and resistant to heat. Plasticized PVC is used for sheeting, coatings for artificial leather and so on.

### 4.3.5 Mixing Behavior

Generally, polymeric mixtures are far less miscible than mixtures of small molecule materials. This effect results from the fact that the driving force for mixing is usually entropy, not interaction energy. Polymeric molecules are much larger and generally have much higher specific volumes than small molecules, so the number of molecules involved in a polymeric mixture is far smaller than the number in a small molecule mixture of equal volume.

The phase behavior of polymer solutions and mixtures is more complex than that of small molecule mixtures. Most small molecule solutions exhibit only an upper critical solution temperature phase transition at which phase separation occurs with cooling, polymer mixtures commonly exhibit a lower critical solution temperature phase transition at which phase separation occurs with cooling.

In dilute solution, the properties of a polymer are characterized by the interaction between the solvent and the polymer. In a good solvent, the polymer appears swollen and occupies a large volume and in a bad or poor solvent, the intermolecular forces dominate and the chain contracts.

### 4.3.6 Polymer Degradation

The susceptibility of a polymer to degrade is dependent upon its structure. Carbon-based polymers are more susceptible to thermal degradation than inorganic polymers and may not be suitable for high-temperature applications.

Polymer degradation can occur due to one or more factors such as: heat, light and chemicals. Degradation affects the tensile strength, color, shape, molecular weight and durability of a polymer material. It is often due to change in polymer chain bonds via hydrolysis, leading to a decrease in the molecular mass of the polymer. Although such changes are not desirable, in some instances, biodegradation and recycling of polymer wastes will help prevent environmental pollution.

Polymer degradation can also be useful in biomedical applications. For example, when a copolymer of polylactic acid and polyglycolic acid is employed in medical stitches, they will slowly degrade after they are applied to a wound.

### 4.3.7 Density

The density or its reciprocal specific volume is a commonly used property for polymeric materials. As with other properties, the specific volume is greatly affected by the temperature and also the processing pressure. Density can be defined as mass per unit volume and expressed as pounds per cubic foot or kilograms per cubic meter. In cellular foams, density is a key factor and also influences the support factor – indentation force deflection (IFD) which is an important marketing tool for bedding and furniture.

### 4.3.8 Fatigue Tests

Dynamic loading of a material that leads to failure after a certain number of cycles is called *fatigue* or *dynamic fatigue*. This factor is important in actual use since a cycle or fluctuating load will cause a plastic component to fail at much lower stresses than it will under monotonic loads.

## 4.4 Environmental Effects

When plastic materials are exposed to the environment, whether the materials are loaded or unloaded components, environment plays a significant role on their properties, affecting their life span, strength, aesthetics and so on. The environment can be a natural one such as rain, hail, solar UV light radiation, extreme temperatures, solvents, oils, detergents or other. Damage in a polymer component due to natural environmental conditions is usually referred to as *weathering*.

### 4.4.1 Water Absorption

While all polymers absorb water to some degree, some are sufficiently hydrophilic that they absorb large quantities of water to significantly affect their performance. Water will cause a polymer to swell, serving as a plasticizer consequently lowering its performance such as electrical and mechanical behavior. Increase in temperature will result in free volume between the molecules, allowing a polymer to absorb more water.

### 4.4.2 Chemical Degradation

Liquid environments can have positive and negative effects on the properties of polymeric materials, while some chemicals or solvents can have detrimental effects on a polymer component. If a polymer component is to be placed or coming into contact with a possibly harmful chemical, it should be checked first. A polymer material that is in a soluble environment is more likely to generate stress cracks and fail. Environmental stress cracking or stress corrosion in a polymer component generally occurs due to the existence of crazes or microcracks.

## 4.5 Viscoelastic Materials

These super materials are generally foams which are four dimensional in relation to properties such as: *density, firmness, time* and *temperature*. Widely used for bedding and furniture applications, the two main basic properties for assessment of quality are density and support factor. For good quality materials, the density should be 4.0–5.0 lbs/cu ft and the support factor – IFD – is greater than 2.0. Since there are many qualities of viscoelastic foams in the marketplace, a simple test to check good quality would be to take a foam sample – 12 in × 3 in × 1 in – and keep it in a freezer. After 1 h, a good viscoelastic foam will turn very hard/stiff, whereas a normal PU foam will remain unchanged.

### 4.5.1 Density

Density is calculated and expressed as weight/volume in pounds per cubic feet or kilograms per cubic meter. Higher densities ensure maximum comfort as well as durability and maintain their physical performance. Typical densities range from 48 to 96 kg/m$^3$.

### 4.5.2 Firmness

Normally, the firmness of viscoelastic materials can be from supersoft to semirigid. This property is achieved by manufacturers by variations in the formulations. If the support factor (IFD) is lower, it will help to increase better weight distribution of a body, thus alleviating back pressure for maximum comfort, but if too low, the material will "bottom out" negating body comfort. The thickness of the material also plays a big part, and generally for bedding applications, thicknesses between 4 in and 8 in are used.

### 4.5.3 Time

This refers to the "recovery time" of a viscoelastic foam. When an object is placed on its surface, the foam "accepts" the weight and makes an indent. When the object is removed, the foam will recover its original shape. A good example is: when a person lies down on a viscoelastic foam surface, the body will sink slowly into the foam, taking the shape of the body, thus enhancing maximum comfort. When the person gets up, the foam will slowly recover and gain its original shape, in this case, a flat surface.

Viscoelastic foams or materials come with different time recovery factors depending on the quality of the foam, the density and the surrounding temperature. This can be adjusted to desired levels during formulation.

### 4.5.4 Temperature

The physical properties of viscoelastic foams can be influenced to a great extent by temperature. A body temperature of a person or the room temperature can affect firmness and recovery rates. Recovery rate is positively correlated to heat, so that increases in temperature will change pliability, firmness and recovery rates. In colder conditions, viscoelastic materials tend to be firmer or even stiff. Ideal conditions for viscoelastic materials to work at peak performance are between 26 °C and 40 °C.

### 4.5.5 Humidity

Viscoelastic foams react to humidity as well as temperature. Visco products tend to soften in more humid conditions. For example, a very pliant foam may feel "slick" or "buttery" as opposed to coarse, depending on the level of humidity. Since in most cases, the viscoelastic material will be covered in cloth or the other, the surface feel is generally not critical in most end applications.

## 4.6 Electrical Properties

In contrast to metals, common polymers are poor electricity conductors. Similar to mechanical properties their electrical properties depend to a great extent on the flexibility of the polymer's molecular blocks. The most commonly used electrical property is the *dielectric coefficient,* also known as the *electric permittivity.* This property describes the ability of a material to store an electric charge.

## 4.7 Optical Properties

This is an important property, especially in automobile and display applications. Since some polymers possess excellent optical properties, they can be easily molded and used to replace transparent materials such as glass. Organic polymers are very good substitutes for inorganic glass as they tend to shatter and be harmful. A good example is windshield glass in automobiles, although the polymer used should be high-impact and scratch-free.

## Bibliography

[1]   Osswald, Tim A., Baur, Erwin, Brinkmann, Sigrid, Oberbach, Karl, Schmachtenberg, Ernst. International Plastics Handbook/The Resource for Plastics Engineers, P. 67–72, 207–212, Hanser, 2006.
[2]   American Chemistry Council, "The Basics: Definition and Properties", https://plastics.ameri canchemistry.com/plastics/Thebasics, 2006.
[3]   Defonseka, Chris. "Practical Guide to Flexible Polyurethane Foams", P. 8, 18, 38–40 Smithers Rapra, 2013.
[4]   Defonseka, Chris. " Practical Guide to Water-Blown Cellular Polymers", P. 3–6, Smithers Rapra, 2016.

# Chapter 5
# Additives for Polymers

## 5.1 What Are Polymer Additives?

Polymer additives can be broadly classified as polymer stabilizers or functional agents. Polymer stabilizers are essential for practical use, since they maintain the inherent properties, color and other characteristics of plastics by preventing the oxidative degradation caused by high temperatures during processing and ultraviolet (UV) exposure during use. Functional additives, on the other hand, are added to further improve the mechanical strength of plastics or impart new properties, such as flexibility and flame retardancy, thereby expanding the range of applications and enhancing their commercial value.

The inherent features of polymer materials are most often not enough to meet the expectations of end applications or to meet specific needs of specialty applications. Thus, polymer processors will seek suitable additives to incorporate to meet these properties. Modifying surface properties of a given plastic or elastomer is among the most difficult goals to achieve. When using additives, processors will also be concerned about compatibility and costs.

Monomers are the basic material from which "plastics" are made. As we know some of the most popular monomers are ethylene, styrene and propylene, and these monomers need small amounts of "additives" to stabilize an increase in their functionality in the first phase of conversion into plastic resins and then again additives are needed when these resins are processed into final products.

As we look for better materials, with costs always a key factor, the demand for plastic material which has more or less replaced traditional materials is expanding. As scientists, chemists and researchers develop newer plastic materials, and the search for better additives for use as single components or in combination also is expanding. Until a few years ago, most additives have been petro-based except for a few mineral-based ones but now in keeping with the policy of moving away from environmentally harmful products, emerging technologies have been enabling resin developers to use bio-based additives effectively.

In practical applications, most plastics contain many polymer additives, each of which is added in small quantities or even as combinations. These additives contribute in various ways, first to produce a processable plastic resin and then new or specialty functionalities as may be required by an end user. Plastic resins inherently lack sufficient stability against heat and light. Because of this, plastics readily oxygenate and degrade when exposed to heat or UV rays. A good example is: when plastics are used in automobiles and exposed to sunlight, they can degrade with discoloring, impair its properties and also impair the durability that is required.

https://doi.org/10.1515/9783110656152-005

In order to prevent this degradation, the addition of small amounts of stabilizers is more effective and economical than modifying the polymer itself. The market has an abundance of stabilizers in a wide range and these can be selected and blended in accordance with the end properties desired of the plastic product. These possibilities have made plastic resins truly very versatile and capable of a wide range of applications. For example, polypropylene (PP) is widely used in many applications but if PP with no incorporated antioxidant is subject to heat at 150 °C, it oxygenates and degrades in 24 h. In contrast, PP containing 0.4% antioxidant does not degrade even after 2,000 h or longer. This means that its lifetime is dramatically improved, while still maintaining the inherent properties and characteristics of the material.

However, in order to expand the range of applications in which polymers can be used, it is not enough for polymers to merely retain their original properties. By blending functionalizing additives into polymers in appropriate amounts, polymers can be given functions, they would otherwise not possess, and the other alternative being modification of the molecular design or polymerization conditions. For example, PP which is translucent will exhibit dramatically improved transparency when 0.2% of a clarifying agent is blended into it. Some other examples of functional additives include flame retardants for flame proofing, plasticizers for added flexibility, antistatic agents to negate the build up of static electricity and nucleating agents, stiffening agents for polymers and thickening agents for plastisols to prevent penetration through the fabric substrate. The recent discovery of the potential of *graphene* as an additive for polymers is quite interesting for enhanced electrical conductance, increase product strength and also in solar substrate coating applications.

Some of the main and common additives used in industry include stabilizers, processing aids, plasticizers, antistatic agents, blowing agents, fillers, coupling agents, antibacterial agents, catalysts, surfactants, lubricants, cell openers, colorants and secondary blowing agents.

## 5.2 Effects of Additives on Polymers

Plastic polymers have chemical reaction properties similar to those of small molecules, though the polymers themselves are larger in size. This means that a range of different factors, including thermal conditions, stress cracking or the diffusion of chemical additives, can alter the molecular structure and thus the fundamental properties of most plastic materials.

Most plastic additives are introduced or incorporated into a compound to produce a specific property or properties, whether to increase processability, foamability and fluidity as in coatings or even as a simple task to change a color or could be a combination to achieve a few desired properties. In the case of copolymers, which are composed of varied and repeating molecular units, each additive incorporated plays a part in the overall chemical makeup. This makes it important to carefully control the amounts and types of additives to be used and compatibility among them may also be a factor as each of them can individually affect the final phase of processing and production.

The author presents later a few important areas where additives are required to achieve desired end properties, including aesthetic values.

### 5.2.1 Plasticizing Agents

While some polymers like rubber are naturally flexible, others such as *lignin* or *cellulose nitrate* are comparatively rigid and cannot be softened by exposure to nonsolvent materials. For this reason, plasticizing compounds may be added to a polymer to reduce its stiffness and increase its processability. Plasticizers function by flowing in between molecules in polymer chains, without altering the polymer volume.

Plasticizers should generally have a solubility level close to that of the polymer itself and multiple plasticizing additives can be used in a single mixture as long as they are compatible with each other and the polymer. When a plasticizer such as dioctylphthalate is used in a polyvinyl chloride (PVC) polymer, it lowers its melt viscosity and increases its light stability, while increasing resistance to oxidizing acids. Plasticizers are widely used in coatings and foaming applications where polymer/copolymers are converted into *plastisols* with other additives. Formulations may include combinations of plasticizers to promote ease of processing and also to achieve desired end properties. When artificial leather is used in automobiles, the inherent plasticizer/plasticizers tend to migrate or "evaporate" due to continued exposure to heat via sunlight (UV action) and protective surface coatings are used to prevent this.

### 5.2.2 Pigments, Dyes and Masterbatches

In general, most plastic resins are colorless, opaque or translucent irrespective of whether they come in the form of pellets, beads, powders, liquids or any other form. Manufacturers of plastic resins and polymers can color them during polymerization to achieve best results of self-color or offer suitable pigments, dyes or masterbatches to producers of plastic products to color them as desired. In all cases, it is important for homogenous mixing to achieve good even color. To assist in this, producers of masterbatches will use the same resin as a carrier or matrix. Since this is an important area of manufacture, especially when aesthetic values are of prime importance, a product manufacturer can work closely with a supplier or directly with the color manufacturer using their laboratory facilities, if needed.

The "color wheel" or "color circle" is a basic tool for combining colors comprising of 12 basic colors. These can be blended as desired to get any final color desired. According to the primary color system, colors are red, yellow and blue. The author suggests that two other colors – black and white – should be added and this forms a more realistic base as these two colors are also "self-colors." These basic colors can be combined to form any color – for example, yellow + blue = green, red + white = pink and red + black = brown. Other variations are also possible with combinations of more than two.

In selecting the correct color product for processing polymers, other than the usual factors includes heat stability, compatibility and durability. When coloring cellular polymers (e.g., polyurethane [PU] foam), for best results the pigments or dyes to be used should be first incorporated into the polyol and mixed before other chemicals are streamed in. Most plastic resin molders like injection, extrusion and blow molders will prefer the colorants in solid form or already self-colored, which naturally will cost more. Cellular polymer producers will want them as liquids or as powders. In the case of expandable polystyrene beads which are generally without color, if colored on a production floor using dyes, the resulting color will only give a "mottled" effect, whereas self-colored beads during polymerization will give full color but these grades will cost more.

### 5.2.3 Stabilizers

All polymers need stabilizers in various forms. Stabilizers can be categorized into four types. The first category is antioxidants, which protect materials during processing and extend their longevity. They are used to prevent polymer degradation which can result in loss of strength, flexibility, thermal stability and color. Antioxidants eliminate oxidation during and after processing, if materials are exposed to an energy source.

The second category of stabilizers is those that help materials to withstand UV light. UV radiation damages the chemical bonds of polymeric materials. Therefore, it is essential to add UV stabilizers to materials exposed to long periods of UV action, especially for products being used outdoors. These stabilizers absorb high-energy UV radiation and then release it at lower energy levels that is less harmful for the polymer. For example, titanium dioxide has a high refractive index that enhances long-term stability and protects against material discoloration.

Very important stabilizers are heat stabilizers, which is the third category. These prevent thermal breakdown of materials and preserve aesthetic values. They eliminate chemical decomposition during processing and in some processing systems, a combination of different heat stabilizers may be used to obtain best results.

The final category is flame retardants. Many polymers are flammable in their pure form and this is a drawback, unless the materials can be stabilized. According to universal building codes, all relevant material in a building industry must be certified for flame retardancy. Flame retardants slow down combustion or create a new situation where less heat is produced and finally dies out. These additives can be easily incorporated into material and will have no impact on the physical properties of materials.

### 5.2.4 Lubricants

Polymers that exhibit difficulty in formability may benefit from lubricants or processing aids that help the material from sticking to metal surfaces or others. Mold release agents in the form of silicone are commonly used as a spray on metal or coated paper for continuous PU foam production operations. Lubricants can take the form of external or internal components, for example when preparing PVC coatings, stearates are added in a formulation.

### 5.2.5 Blowing Agents

In the production of cellular polymers, also called foamed polymers, blowing agents are used. These can take the form of primary or secondary blowing agents, the latter being required when very light weight densities are desired. For example, PU foam manufacturers use a primary blowing agent and for lower densities, a secondary blowing agent like methylene chloride is added. It is interesting to note that due to environmental concerns and in an effort to move away from petro-based blowing agents, emerging technologies are coming up with water-blown cellular polymers.

### 5.2.6 Catalysts

A catalyst is a substance that alters the rate of a chemical reaction but remains unchanged. A "positive" catalyst accelerates a chemical reaction, while a "negative" catalyst slows down a reaction. Catalysts can even take the form of a "system" to achieve balance between several chemical components. A good example is the use of an amine catalyst and a tin catalyst to control gas formation, cream time, rise time, gel time and final cure of the foam with strong cell walls in a PU foam formulation. Organometallic compounds based on mercury, lead, tin, bismuth and zinc can be used for cellular foams. In PU production, the gelling process between the isocyanate and the polyol is best promoted by tin catalysts.

## 5.3 Influence of Additives During Processing

Whatever method is used for processing polymers, for ease of processing and manufacture of good quality products, processing aids are recommended and, in most cases, are a must. The choice of processing aids may differ from process to process and also each manufacturer or processor may have their own preferences. The choices may also depend on the machinery used in a particular process and the design of molds or equipment used.

Since most polymers are processed at high temperatures, prevention of polymer degradation is of utmost importance. A few common processing problems are presented further:

### 5.3.1 Discoloration and Degradation

This can be caused by contamination of polymer degradation due to excessive heat. Degradation takes the form of polymer discoloration such as black spots, pinholes, traces of volatilized matter (smoking) or deterioration of the properties of the resulting molded article.

The main types of degradation are *thermal, UV, chemical, mechanical, radiation damage* and *biological degradation* with some of these types occurring together. For example, polyethylene is prone to thermos-oxidation degradation. This is a combination of thermal and chemical degradation caused by the presence of oxygen.

One solution is to add antioxidants using the synergic effect produced by the combination of a phenolic antioxidant with a phosphorus antioxidant. Also, lower the molding temperature and shorten the period during which time the polymer is exposed to high temperatures.

### 5.3.2 Improving Molding Processability

In injection molding, for example, meeting the following requirements will greatly help the process:
(1) High resistance to thermal decomposition
(2) Good melt flow
(3) High injection speed to allow short cooling time
(4) Efficient cooling system – short dwelling time
(5) Good part release from mold

For complex and thin parts, especially, where productions are large volume, short molding cycles are required, primarily for productivity and reduced molding costs. For these reasons, it is necessary to assist the molding cycle by the use of additives such as antioxidants, thermal stabilizers, lubricants, nucleating agents and release agents.

### 5.3.3 Improving Durability

For most plastic products, durability is an important property. In order to enhance the long-term reliability of plastic particularly against thermal and UV degradation, suitable additives have to be used. When plastics are exposed to high temperatures, they thermally decompose over time due to the influence of heat and oxygen and may undergo change in properties.

Since the energy of UV light is higher than the molecular bond energy of a polymer, when molded articles are exposed to UV light, it degrades over time, starting from the surface and may undergo chalking, discoloration and deterioration of properties. Adding of an UV absorber or a hindered amine light stabilizer will prevent deterioration, especially from weatherability.

For each additive added to the polymer, although larger amounts provide more effective improvement of the relevant properties, excess could result in negative effects on other properties. Ideal would be to use the minimum amounts of each additive required. In crystalline polymers, additives are dispersed mainly into a polymer's noncrystalline areas and are therefore prone to emerge onto the surface of the molded article at the glass transition temperature or higher. One method of preventing this problem is to use additives of high molecular weight.

When selecting additives, it is important to give careful consideration to some of the basics as follows:
(1) Compatibility with the polymer
(2) Volatility
(3) Reactivity
(4) Color properties
(5) Cost factors

### 5.3.4 Use of Lubricants

In plastic molding processes, the processability can be greatly improved by lowering the melt viscosity of the mass which will allow fluidity and good flow. In single cavity molding, it may be more or less straight forward but in multicavity operations, good flow is essential to fill all cavities fully. The ideal would be to use a lubricant closely similar in structure to the polymer and compatible with it. Lubricants can be either internal or external.

External lubricants are used to improve lubricity between the material and the processing machines where good flow of the hot melt is desired and then free flow into molds. At de-molding times, easy release is essential, especially important in multiple automolding processes. External lubricants should have low compatibility with the polymer as they function on the surface with easy release and improved surface finish. A good example of an external lubricant is a release agent silicone spray for nonadherence of a molded product. Another example is the use of sili-cone-coated paper forming a trough during continuous PU foaming systems.

Internal lubricant's compatibility with the polymer will have good flow, be able to blend into the interior and enter and fill the spaces between polymer chains, thereby lowering melt viscosity. This will help to form a homogenous melt and also lower the torque resistance of a screw during plasticization. Plasticizers are com-mon internal lubricants used in polymers which soften a polymer by entering spaces between chains and weaken their intermolecular forces, thereby improving fluidity.

## 5.4 Functions of Some Additives

Additives can be used singly or in combination of two or as a system. Perhaps the most important aspect, other than cost, would be compatibility with the polymer and also compatibility among them if used as a system. The following shows the main functions of some of the commonly used additives:

### 5.4.1 Antioxidants

Antioxidants help prevent "oxidation," meaning the polymer reacting with oxygen. Oxidation can cause discoloration, surface cracks, loss of impact strength and elon-gation. Antioxidants prevent thermal oxidation reactions when plastics are proc-essed at high temperatures and light-assisted oxidation when plastics are exposed to UV light.

### 5.4.2 Antistatic Agents

Prevents build up of static electric charges. Plastics are generally good insulators, so they have the capacity to accumulate static charges on the surface which will hinder processing procedures and can also have issues with hygiene and aesthetics.

### 5.4.3 Biodegradable Plasticizers

As volumes of postconsumer plastic wastes grow, it is always a problem to get rid of it without creating environmental concerns. Plastics have an inherent property of very slow decomposition and several additives are being used to make it faster or more effective. Biodegradable plasticizers are one of the additives that can be used for plastics to soften and make it more flexible and thus enhance degradability.

### 5.4.4 Blowing Agents

Forms gases in resins to produce cellular foamed plastic materials. Minute amount of blowing agent incorporated in a resin during polymerization and later agitated under heat will form a gas which will influence a polymer matrix to form a foamed material. A good example is expandable polystyrene (PS) resins in the form of beads. Each bead will contain a minute amount of pentane inert gas incorporated during polymerization and will "foam" during the first stage of "pre-expansion."

### 5.4.5 External Lubricants

Generally applied to mold surfaces for easy release of molded part. Excessive application will leave a "haze" or slightly whitish appearance on the molded part surface. In some cases, they may be applied directly to the material. Applying release agents to mold surfaces may be important especially in multimold auto processing.

### 5.4.6 Fillers/Extenders

Incorporation of natural substances like calcium carbonate and others help to increase bulk, lower costs and alter density of a material. Mineral-based fillers/extenders will increase the overall "bulk" and also make foams heavier. In PU foams, this factor is important for bedding and furniture applications where the indentation force deflection factor is required to be >2.0.

### 5.4.7 Flame Retardants

These prevent ignition or spread of flames in plastic materials. Plastic material in various forms is used in industry, building construction, transport applications and so on, and each industry will have its own regulations for materials. Manufacturers have to ensure that any plastic materials made for these applications will comply with the local regulatory requirements.

### 5.4.8 Heat Stabilizers

Since in most cases, processing temperatures is above 180 °C, polymers will need the addition of heat stabilizers to prevent decomposition during processing.

### 5.4.9 Impact Modifiers

Enables plastic products to absorb shocks and resist impact without cracking. Particularly relevant for polymers such as PVC, PS and PP.

### 5.4.10 Light Stabilizers

To protect reactions in plastics which cause undesirable chemical material degradation from exposure to UV light.

### 5.4.11 Coloring Agents

These can be pigments, dyes and masterbatches in the form of liquids, solids, pastes and powders which are needed to achieve any particular color desired. Excessive amounts will "migrate" to the surface and display a condition called *splay*.

### 5.4.12 Reinforcements

Additives used to reinforce or improve tensile strength, flexural strength and stiffness of the plastic material. Fiber-based additives are common, while rice hulls (20% silica) and bamboo fibers will make excellent reinforcing constituents, particularly in polymer composites.

Daikin Global offers a range of high-performance lubricants, additives, antidripping modifiers and processing aids as compounded fluoropolymers. Blended in conventional polymers, they improve the final plastic product or elastomer performance, increasing processing ease and also increasing production yields. Daikin range of processing aids and additives have been developed based on fluorine technology and these will provide solutions for conventional plastics such as polyamide, polyacetal, polyester and polycarbonate.

Their additives range particularly suit for compounding and polymer processors looking for
- low friction;
- wear resistance;
- antidripping.

They are also suitable for films and cable producers seeking to
- reduce melt fracture;
- improve gloss;

- reduced gels;
- reduce die build up;
- faster color changes;
- reduced die pressure.

Combined with fluorocarbon or silicone oils, polytetrafluoroethylene micropowders participate in the formulation of high-performance lubricants and greases. They have a wide operating temperature range, very low friction and a much longer usage time than standard lubricants and also offer cost optimization and production efficiency.

# Bibliography

[1]  Adeka, "Introduction of Polymer Additives/Chemical Products", www.adeka.co.jp/en/chemi cal/products/plastics/knowledge_0.4.html-.
[2]  Osswald, Tim A., Baur, Erwin, Brinkmann, Sigrid, Oberbach, Karl, Schmachtenberg, Ernst. International Plastics Handbook, Hanser Publication, 2016.
[3]  Defonseka, Chris. "Practical Guide to Flexible Polyurethane Foams", Smithers Rapra, 2013.
[4]  British Plastics Federation, "Plastics Additives", www.bpf.co.uk/plastipedia/additives.
[5]  Daikin Global, "Polymer Additives", www.daikinglobal.com.

# Chapter 6
# Useful Data for Processing Polymers

## 6.1 Important Parameters

Processing polymer resins can be both exciting and challenging as the uses and applications will range from simple consumer products to high-tech components. For example, when considering making plastic products for industrial use, automobiles and even for space travel, both resin manufacturers and polymer processors will no doubt face real challenges with the ones on the processors, who will be responsible for the quality, durability and aesthetic values and so on of components.

In order to face and overcome these challenges, a good basic knowledge of some of the important aspects of processing polymers will be of great help. This chapter will present valuable data covering some of the important areas of processing which will contribute to a large extent in order to process and manufacture cost-effective quality products.

## 6.2 Coloring of Polymers

The aesthetic values of any plastic component or product is as important as quality. Polymer resins are generally available to processors in their natural state meaning either colorless or transparent or translucent. Depending on customer requirements, these resins have to be "colored," inclusive of black color. In some cases, such as low-end applications, color may not be important or desired.

Colorants are available as pigments, dyes and masterbatches in the form of powders, liquids, blends or others, like metallic colors or specialty compounds. There are low-end and high-end colorants with cost variations. Two of the cost-effective common colorants are titanium dioxide (white) and carbon black (black). In processing polymers, it is important to achieve a good homogenous mass with even color distribution to prevent flaws on the surface of a molded component or product.

### 6.2.1 Introduction

Although synthetic polymers are man-made, it is not unreasonable to think that they are in harmony with nature. In the world of plastics, color plays a very important role and processors must have a thorough knowledge of the various possibilities available. To aid them there are many training and testing facilities available with color suppliers and also independent organizations. An in-house small laboratory on a production floor of a polymer processor will also, no doubt, be of tremendous value.

https://doi.org/10.1515/9783110656152-006

Over the years, polymeric composites also have found applications in various fields such as consumer, automobile, building construction, transportation and space travel.. Coloring of single or blends of polymers may be considered fairly straightforward but composites need more attention and sound technology, especially where they are exposed to the elements. Here, in addition to the basic knowledge of coloring, a good knowledge of selecting, blending and perhaps the application of protective coatings will be an added advantage.

Colorants can be classified according to their tone, pigment class, color index, opacity, light-fastness, tinting strength and physiological/chemical properties. Unlike pigments, colorants are soluble in plastics. Manufacturers of colorants try to maintain equilibria between natural and synthetic materials and other than the many standard ones available, most manufacturers offer custom-made services to meet any type of color and properties required. With advances in technology of masterbatch coloring, additive masterbatches contain desired additive agents to counter UV effects, flame retardants, moisture control and others.

### 6.2.2 Theory of Colors

The basic purposes of coloring polymer products can be for aesthetic value, setting a mood, attracting attention, making a statement or for product protection. Colors tend to energize or cool down products to blend with its surroundings and thus be more acceptable. By choosing the right color blends, one can create an ambiance of elegance, warmth or tranquility. Particularly in consumer products, interior décor in buildings and other applications, color can be a powerful design element if one knows how to use colors effectively.

The origin of colors is sunlight, which has thousands of colors but to our eyes only a few are visible. Most of us are familiar with the range of colors as shown in a spectrum: violet, indigo, blue, green, yellow, orange and red. These are the colors we can perceive. Colors affect us in different ways, both mentally and physically. For example, a strong red color may raise one's blood pressure, whereas a blue color will have a calming effect. Being able to offer aesthetically pleasing polymer and composite products with color consciousness in harmony to end users, especially in the building construction, automobile, consumer products and furniture industries, can help achieve spectacular results.

### 6.2.3 The Color Wheel

The color wheel or color circle is a basic tool for combining colors. According to reports, the first color diagram was designed by Sir Isaac Newton in 1666. The color wheel is designed so that virtually any color one may pick from it will be compatible with another color. Over the years many variations of this basic design has been made but the most popular one is the original wheel of 12 colors based on red, yellow and blue (RYB). Several color combinations are considered pleasing and are called *color harmonies or color chords,* consisting of two or more colors. Colors are categorized broadly into *primary, secondary and tertiary.*

### 6.2.4 Primary Colors

In the RYB color model, the primary colors are red, yellow and blue. However, I would like to add two more important colors *black and white*, making a base of five colors instead of three. Primary colors mean that these colors cannot be achieved by the mixing of any other colors. With this base of five colors, it is possible to achieve any color.

### 6.2.5 Secondary Colors

Secondary colors, for example, green, orange, pink, purple and gray are obtained by mixing:

    Green: yellow + blue
    Orange: red + yellow
    Pink: red + white
    Purple: red + blue
    Gray: black + white

### 6.2.6 Tertiary Colors

These colors are obtained by mixing the primary and secondary colors.

### 6.2.7 Warm and Cool Colors

The color circle can be divided into warm and cool colors. These colors become important for materials made with polymers and composites for use in interior décor, facades, paneling, ceilings, furniture and in automobile interiors. *Warm colors* are vivid and energetic and tend to "expand" in space, whereas *cool colors* exude an impression of calm and create a soothing effect. White, black and gray are considered as neutral colors.

### 6.2.8 Tints, Shades and Tones

Tints, shades and tones are often used incorrectly, although they are simple color concepts. If a color is made lighter by adding white, the result is called a *tint*. If black is added, the darker version is called a *shade*. If gray is added, the result is a *tone*.

## 6.3 Purging Colored Material from Production Machinery

It is well and good that polymer processors have to use color additives to produce aesthetically pleasing products. Whatever process is used, it is important to be able to change color when needed, for example, at the end of a particular production run and when the machine is being made ready for another product with a different color. If uncolored resin have been used, the changeover may be a bit easier with purging of the leftover material in the barrels.

Since most polymers are not originally colored, it will take calculated addition of colorants such as masterbatches and colored pellets to achieve what is desired. It is imperative that this colored material is fully taken out of a molding machine so they do not linger too long and contaminate new production runs. Thus, for molders, purging becomes an important part of any process. With a wide range of purging compounds available in the market, molders have to decide which one will be best to use. It may be that out of these commercially available purging compounds (CPCs), a molder may require more than one compound.

Switching a molding machine directly from one color to another without proper purging will result in colors from a previous run mixing up with the new color and coming out in the form of streaks, with many products ending up as rejects. As a general rule, the choice of a CPC is more or less determined by the type of processing machine, the material used and the processing temperatures. Generally, for mechanical purging compounds that tend to work based on material affinity and viscosity differences, using the machine to do the work may be considered the best for injection molding. Chemical purges that tend to have foaming agents in them allowing the purge to expand and flow into the low-flow areas or poorly designed areas are probably the best for extrusion machines.

When it comes to color changeovers, molders have a lot of options such as Dyna-Purge A for PP from Shuman Plastics, Inc., ASACLEAN E by Sun Plastech, a styrenic-based CPC for both injection molding and extrusion and Ultra Purge PO from Chem-Trend to name a few.

### 6.3.1 Masterbatches

Most plastic resin processors use masterbatches which are concentrates of pigments dispersed into a polymer resin which acts as a carrier. The ideal would be for the carrier to be the same polymer or at least a polymer compatible with the polymer resins to be colored. A processor can be guided by the data supplied by a master-batch supplier as for predefined ratios to be used, processing temperatures and other relevant data. Depending on the carrier selected, the same masterbatch could be used to color several different polymer resins. Since most manufacturers of mas-terbatches have efficient in-house laboratories and services, when special colors or blends are needed, they will oblige.

### 6.3.2 Universal Masterbatches

Universal masterbatches are also available from some suppliers, particularly from RTP Company (USA) under the brand name *Unicolor*. These coloring systems are an innovation to coloring when multiple polymers are being used in a processing plant, including engineered polymer resins. These products can be used with low ratios such as 1–2%, depending on the polymer resin and do not require drying be-fore processing. These can be extremely useful when making polymeric composites with different polymer matrices such as with pure resins, recycled resins or even combinations of both.

Some of the basic colors offered by suppliers are white, beige, terracotta, brown, yellow, amber, red, green blue and light-gray. Singly, combinations or variations of these colors can be used when wood-grain veneers are desired for solid wood-like composites when they are being made. These products are compatible with polymers such as polyethylenes, polypropylene, polystyrene, polyvinyl chloride, acrylonitrile-butadiene-styrene and polycarbonate.

### 6.3.3 Liquid Colorants

Different systems may be available in the marketplace but the basic liquid colorant systems will comprise of at least one pigment and one dye for coloring cellulose polymer composites. Composites with wood-grain effects have been made for a long time with combinations of wood and plastics. Emerging technologies have found that composites with rice hulls will have superior properties to those made from wood chips and wood wastes. The high content of silica in the rice hulls en-hances these composites.

A liquid colorant system can contain 0.5–10% by weight of dye based on the pigment. This pigment can consist of an organic or inorganic pigment plus at least one dye, a dispersant and water or a mixture of water and a water-retaining agent. Preferably, the pigments should be of finely divided particles, typically of average size of 0.1–0.5 μm.

### 6.3.3.1 Special Effects Liquid Colorants

Riverdale Global – USA offers a new special effects liquid colors designed to help manufacturers of packaging and consumer products with enhanced shelf appeal and with added value in cost-effectiveness as compared to standard masterbatches.

The new range of specialty colors includes basic colors which provide a pearlescent effect. For example, *Deep Pearl* uses particles that provide a standard pearlescent effect at lower let-downs; *Transparent Pearl* is used at 0.5% loading in clear resins and exhibits a glitter effect; *Blast* provides a super-bright pearlescent effect, while minimizing flow lines; *Splash* combines bright specialty pigments and pearlescent particles to give a super-bright effect; *Metal* provides a smooth, flat surface with a metallic sheen; and *Metal Expression* has larger particle sizes that create a glittery metallic effect.

### 6.3.4 Precolored Resins

Naturally, precolored resins will cost more than using standard resins for coloring on a factory floor. However, polymer processors without much experience in handling colorants may opt to use precolored resins for ease of use or a specialty job may require the use of these resins to ensure good color distribution and quality. When processing engineering resins to make high-end components, precolored resins may be an advantage to obtain full benefits of color, since pigments are completely polymerized into a resin and molders can use them as supplied.

### 6.3.5 Cube Blends

Cube blends are color systems that feature a masterbatch, dry blended with natural polymer that are ready to use by a processor. This will benefit molders who lack mixing or blending equipment on a factory floor. The use of cube blends will eliminate the risks associated with self-blending that can cause variations in color.

### 6.3.6 Colored Specialty Compounds

Colors enhance and add value to custom engineered components processed to meet specific physical requirements. When dealing with production of high-tech components where color and other properties are important, processors can contact any of the companies who deal with colorants, additives and resin blends to enhance success. It is important that whatever color or additives are used, they should not alter the physical properties of the components, in any way. In some cases, customers may demand an exact color match and here specialty compound suppliers can help in a great way. Some of the color requirements for finished products could be matt finish, high gloss, cosmetic appearance and metallic effects.

## 6.4 Electrical Power Calculations

The supply systems of an electropower supply are the generating stations, the transmission lines, the substations and the distribution networks. While the generators produce high voltage electrical power, the overhead transmission lines moves it to the regional main stations and to substations, where the high voltages are "stepped-down" to workable loads and supplied to industrial, commercial and residential purposes. Some relevant keywords are amperes, voltages, current, single phase, three phase, step-down transformers, circuit breakers and also adequate gauges to be used. This is important as the use of marginal or under specification electrical wiring cables, mainly for purposes of cost-saving will create possible fire hazards due to heating of the conductors. A three-phase circuit is simply a combination of three separate single-phase circuits but conventionally connected. A well-designed industrial plant will have one or more power control boards inside the factory with circuit breakers for protection due to emergencies and also to counter power surges which can be positive or negative and harm motors and equipment.

An industrial manufacturing unit will basically require electrical power to run its machinery, equipment and for lighting. General installation supply will be three-phase power from which both single-phase and three-phase power can be drawn depending on the requirements of motors and equipment. Here, the author presents some guidelines for calculating electrical power requirements.

Whatever machinery and equipment is used, the suppliers will indicate the horsepower (HP) of each motor or device, the voltage and other information in their specifications sheets. For example, a motor will be specified as a single-phase 1 HP motor or as a three-phase 5 HP motor with an indication of the voltage, which can range from 110/120 V and 230/240 V for single phase and 380/440 V for three phase. A basic manufacturing plant will need electrical power for motors, equipment, lighting, air conditioning water supplies and so on. Electrical power is generally expressed as kilowatts (kW) and electrical consumption costs are based on a basic unit taken as kWh – the quantity of electrical power consumed in 1 h.

The following calculations are based with an emphasis on power requirements for machinery and equipment and the same calculations can be used to work out the requirements for others as well. When installing a power supply, an additional

40–50% should be added on the basis that all motors/machinery will be started up at the same time. Once the overall power requirements have been worked out, it is always prudent to give consideration for possible expansion programs.

Calculating amperes ($I$) when HP is known:

- Single phase = (HP × 746) ÷ ($E$ × Eff × pf)
- Three phase = (HP × 746) ÷ (1.73 × $E$ × Eff × pf)

where $I$ represents current in amperes; $E$ is the voltage in volts (120 V used in calculations); Eff is the efficiency as a decimal; Pf is the power factor as a decimal; and HP is the horsepower taken as 746.

The following is a simple example calculating power consumption for a small plant:

Example: A small plastic injection molding plant has
- one 1 HP motor;
- one 3 HP motor;
- 15 lights (each 100 W).

Then, estimated power consumption = (1 × 746) + (3 × 746) + (15 × 100) W
$$= 4,484 \text{ W} = 4.484 \text{ kW}$$

Therefore, estimated power costs will be around 4.5 units per hour.

Power consumption, if all work for 8 h = 4.484 × 8 = 35,872 kWh

Using formula, W = volts × amperes × power factor

$$4484 = 440 \times \text{amperes} \times 0.8$$
$$\text{Amperes} = 12.74 \text{ per phase}$$

Therefore, this plant can have a three-phase 440 V/50 Hz wiring system to suit 15 amp or 30 amp (better) with a three-phase 440 V circuit breaker trip switch with individual start/stop switches at point of power entry. If a processing plant is located in an area experiencing frequent power failures, it may be prudent to have a standby generator, depending on viability.

## 6.5 Compressed Air

Compressed air is a valuable commodity for any manufacturing plant. A small plastic processor may be able to manage with a single air compressor but medium to large size operations will require large volumes of compressed air and a good distribution system inside the plant. Generally, this would mean a "compressor room" where several large compressors are inter connected and fed into a large accumulator tank from which a strategic pipe line system can draw.

As a general rule, the standard compression ratio is 5:1, meaning that a 5 HP air compressor will yield about 1 HP of compressed air for industrial purposes.

Compressed air generates heat and water concentrations with possible other contaminants and these have to be removed by using water traps, valves and other devices to ensure a clean and dry air supply at strategic points. The two important factors in an air supply are (1) pressure and (2) volume. However, air supply lines are bound to have a "drop" from the point of supply to the point of draw and if this is around 8% it is acceptable by industrial standards. Therefore, for maximum efficiency this factor can be considered when calculating the overall requirement for the plant.

## 6.6 Water Requirements

An important requirement for any operation, be it small or large. There are many systems which can be installed and used. A small factory may have a ground-well with an overhead tank/tanks to supply water for different uses. Larger plants will probably take the water supply from the main lines and may have a recirculation system. Some of the uses for water are for drinking, washrooms, flushing machines, cooling molds, showers where chemical spills are a possibility, steam boilers and eye wash stations.

Water will be required in large volumes in cases of fires also. Water usually contains impurities and depending on the end use, will have to be treated to remove the calcium content when used for steam boilers as it will otherwise corrode the steam tubes. It would be a good idea to use treated water for cooling molds wherever possible. To ensure steady and continuous supplies of water, it may be prudent to have large storage tanks to draw from. However, these tanks must have sufficient valves at the bottom to flush out the "silt" accumulating over long periods.

## 6.7 Safety Factors

Irrespective of the size of a processing plant, safety factors are very important. Safety requirements should start with the designing of a plant and two important aspects are easy exits for emergencies as per regulations and good ventilation. Since the subject here is polymer processing, an added dimension should be considered as the materials are based on chemicals. Standard processing of polymer resins via injection molding, extrusion, compression molding, blow molding and so on can have standard safety systems but when it comes to coatings like artificial leather, polyurethane (PU) foams or others where the chemicals are highly toxic and corrosive, special handling has to be implemented. Where toxic chemicals are used on a production floor, the air quality is important and periodic checking (preferably by outside experts) should be carried out to ensure the air quality is within allowable standard limits.

Safety should start with the storage of raw materials, their movement inside the plant, machinery safety, safety of personnel and safety of the plant as a whole. Some of the important items for safety systems start with air quality, water sprinklers, fire alarms, fire extinguishers, spill-management procedure, professional assistance and training for key personnel, emergency showers and eye-wash stations (where applicable), protective wear, appropriate waste disposal methods, safety goggles, safety shoes and others. No-smoking signs and safety posters at strategic points inside the plant will also help to remind all personnel of a safety-first policy. When dealing with hazardous chemicals, for example, in PU manufactures, provision must be made for fire and spills with correct and effective methods to prevent and also deal with such situations, if it happens.

Periodic meetings, fire and spill management containment and effective machinery maintenance will go a long way toward maximizing safety. Depending on the polymer processing operation, some of these relevant safety features should be implemented.

## 6.8 Quality Control

In any manufacturing or processing operation, good quality control (QC) is an important aspect in order to make good quality products. This section provides basic guidelines and explanation of a common QC system used in polymer processing-statistical process control (SPC) – as specified by the International Organization for Standardization (ISO) for the ISO 9000 series for general purpose and QS 9000, which is a special standard for the automobile industry.

Depending on the products and requirements of customers and or the marketplace, a processor will choose a system such as ISO 9000 and ISO 9002. Once the SPC system is implemented and running, constant evaluation, monitoring and corrective action is needed. Standard machines will have an "open-loop" system where periodic corrections may be needed to keep the machine operation "in-line" or within preset parameters. Advanced machinery will have "closed-loop" systems, where once the parameters or limits are preset, the machine will automatically adjust itself.

SPC systems start with the raw materials, the processing stage and a final check for quality by a floor quality inspector or supervisor. General color codes used are green (acceptable), yellow (on hold) and red (rejects). It is important to understand that quality is not the responsibility of just one person (such as an operator for example) or a section but needs the cooperation of all. SPC uses two basic control charts called X–R chart and P–chart.

### 6.8.1 X–R Control Chart

When plastic parts are made, there are several important aspects that combine together to produce a quality product. In most cases, these aspects are governed by predetermined tolerances as it is virtually impossible to make the same part to an exact specification. This becomes more critical when components are for industrial or more advanced uses such as automobile or even space travel. Thus, an X–R control chart becomes very valuable in "controlling" a process within preset tolerances.

An X–R chart is a graphical representation of the quality of a particular feature of a part or product such as weight, density, dimensions or other generated by a process. This chart set up in a processing machine tells the operator "how well" the process is going, that is, during operation, whether each part being made is within the preset limits. This control chart will have a median (ideal), an upper control limit (UCL) and a lower control limit (LCL). The information provided at any given time, if a process is out of limits will tell an operator to call a technician or supervisor who will make the necessary adjustments to bring the process back into control.

An X–R control chart will have four basic sections:

Section 1 in the title box tells us:
– name of customer
– product identity
– what feature is being measured
– how often it is done

Section 2 is a graph grid with the median, UCL and LCL.

Section 3 shows the areas where an operator will fill in the following:
– date
– time
– measured values
– sum total
– averages
– range
– initials

Section 4 is a graph for plotting the range between readings.

### 6.8.2 P–Chart

A P-chart is used for attribute data. This exercise is to determine whether a part is acceptable, on hold or rejected. Here, the color code system of green, yellow and red can be used for the holding bins.

Every process has some kind of variation and it may be true to say that in all probability, no two parts may exactly be the same but can be accepted if they are within the predetermined tolerances. These variations can occur due to many reasons such as machine performance, variations in the hot polymer melt, voltage fluctuations, variations in injection pressure and poor mold design.

Once a processor has decided on the ISO system like ISO 9000, 9001 and 9002 to be used, the correct paper work should be prepared in keeping with the specifications agreed on. We must keep in mind that several processing machines may be involved, with different products and different specifications. Once an ISO system has been implemented and running, there will be periodic inspections and checks by representatives of the quality management systems of that province or country who set it up to see if ISO system is working effectively and correctly for continued certification.

### 6.8.3 An SPC System in Operation – *An Example*

Take the case of a medium size injection molding plant. Three molding machines of different sizes and supporting equipment are installed. They are identified by their clamping force (e.g., 300 tons, 500 tons or other) and the shot size. They can be electrical, hydraulic, hybrids or other systems with manual, semiauto or fully automatic operation. For manual or semiauto, an "open-loop" system will be installed needing manual re-setting when a process is out of control, whereas auto operations will have "closed-loop" systems to record readings and will automatically correct the process to preset tolerances when out of control.

The management decides to implement an SPC system such as ISO 9000 or ISO 9002, suitable for the parts being made and as required by a customer.

The management consults a quality management organization affiliated to the main ISO body and sets up a QC department with a quality controller, QC inspectors and trains the operators in the new system to be implemented. Each molding station will have a SPC screen mounted on a stand or on the machine itself, where an operator can take periodic readings. A SPC chart, a job card, a logbook and a sample of the part to be made will complete the basics.

In keeping with the specifications given in the job by the QC department at (e.g., record weight via cushion readings at one hour intervals), a technician will start the machine and purge the hot melt through the injection nozzle and then set up a "cushion" (distance between the screw tip end and the injection nozzle opening). This will control the amount of material injected and equal to the weight of the part to be made which can vary during the processing operation due to voltage fluctuations, hot melt density variations, flow deficiencies, especially in multiple cavity molds and so on.

When the molding process is in progress, the operator will note the cushion readings at specified intervals and record them in the SPC chart. The operator will also calculate the range between readings and enter it on the bottom half of the chart. So long as the readings are within the UCL and LCL with median line as ideal, the process is said to be in control. If any of the readings are out of the limits, the operator will alert a technician who will check it and correct the process back into control and also record the corrective action taken at the back of the SPC chart for evaluation later. The operator will also alert a technician if defective parts are being made such as parts with "splay," flash, short or shrinkage.

The productions will then be moved to a holding area for a physical check by a QC operator or a QC inspector where parts will be sorted into bins – green (acceptable), yellow (on hold) and red (rejects) before packaging. Before moving to a warehouse, the finished packs must have a green sticker in addition to the standard packing list. Here, a P-chart will be used for analysis and corrective action, if necessary. A sample board of defective parts placed in a strategic place will greatly help the operators.

## 6.9 Preventive Maintenance for Machinery and Equipment

For efficient processing on a plant floor, assuming that the machinery and equipment have been properly installed with the required safety feature for safe operation, rather than carrying out repairs whenever a breakdown occurs, it is prudent to set up a *preventive maintenance program* which will minimize breakdowns, thus minimizing downtime that can be costly.

In general terms, preventive maintenance comprises care and servicing of organizational assets to maintain them in satisfactory operational condition through systematic inspections, observations in order to detect and correct defaults before a total breakdown occurs. The main purpose of a successful preventive maintenance program other than maintaining them in good order is also to extend the life span of machinery, equipment and tools by predicting possible failures. This is especially important in a plastic processing plant where molds are used which are very costly to replace. Common problems are damage, corrosion, wearing of any cavity protection coatings and improper postmolding storage.

A good preventive plan starts with the suppliers of machinery and equipment who will provide a general minimal servicing needs but a more enhanced plan can be worked out by the management of a plant in keeping with preplanned objectives. All maintenance operations included in the overall plan should fulfill three important areas such as applicability, efficiency and profitability. An operation is applicable if it can be implemented, and it is efficient if it significantly reduces downtime due to failure rates and it is profitable if it improves production. An effective maintenance plan must include spare parts, lubricants, correct tools, knowledgeable personnel, logbooks at each molding station and sound feedback system. Large processing plants may even invest in and install maintenance software for efficiency and to minimize downtime.

The degree or in-depth levels of maintenance will depend on the complexity of operations on a processing floor which could range from standard to very complex production operations. It can even be considered as a lean manufacturing tool – total productive maintenance embracing the Japanese concept of 5S – *seiri* (elimination of unnecessary work), *seiton* (methods), *seiso* (cleanliness), *seiketsu* (control) and *shitsuke* (discipline). Lean manufacturing as conceived by the Japanese basically is the elimination of waste in all actions/sections of an operation for enhanced profitability.

## 6.10 Key Factors for a Processing Plant

For efficient processing some of the recommended key factors on a production floor are
- Proper training for all floor personnel procedures, machinery, equipment, and QC system.
- Documentation, start-up and shut-down procedures of machineries for technicians.
- Posters: motivation, safety, warnings, hazards, accidents and no-smoking.
- Maintenance: regular routine and preventive system.
- Quality assurance program and constant analysis of customer feedback.
- Product development: quality improvement, new products and constant process improvement.
- Workforce: good morale, periodic meetings/training and alert any new procedures.
- Good ventilation and air quality, especially where chemicals are used.
- Raw materials: proper storage, handling and hazards prevention.
- Efficient plant layout for good production flow.
- For large plants: regular meetings between managers, supervisors, lead-hands and others.
- Job rotation: operator rotation between hard/easy jobs.
- Procedures during accidents, fires or other emergencies.
- Good management: communication, motivation, evaluation, and incentives.

## 6.11 In-House Laboratory

It is an advantage to have a small in-house laboratory, even a small one, depending on the size of the processing plant. This can also serve as a combined center for QC operations. Minimum facilities should include measurements, weighing, density checks, QC operations, records and master logbooks for each machine.

## 6.12 In-House Workshop

This will help greatly in the smooth operation of a processing plant. Basics would be the availability of a mechanic and electrician. It is important to carry relevant spare parts and other items needed to minimize production downtime. In addition to routine maintenance of machinery and equipment, a plant may also decide on a preventive maintenance program and an in-house workshop is more or less a necessity.

Preventive maintenance programs, plastic waste granulators and shredders, efficient spill-management training, first-aid facilities, fire extinguishers and perhaps a power generator where possible would greatly enhance efficient production.

## 6.13 Granulators/Shredders for Recycling Plastic Wastes

No plastics processor actually wants to generate wastes on a production floor but this is an inherent aspect of processing plastics. Each organization may have their own standards but generally 4–5% wastes are considered a norm, while the foam industry will generate much higher rates, such as 15–20%. However, they solve this problem by having suitable machinery for size reduction and using either an adhesive/compression or steam/compression process to produce large blocks of compressed foam which are cut into slabs and sheets for use as mattress bases and carpet underlay, respectively.

In plastics processing, wastes are generated at start-up and shutdown time with bad parts adding to the volume. Wastes from purging machines also adds up. Some processors may have small granulators by the side of injection molding, blow molding, extrusion, thermoforming machines and so on. So long as the reduced size is right and there is no contamination with foreign matter, the granulated material can go back directly. All processors are always on the lookout for the latest machinery which could enhance their productivity. Here are some of the latest offerings from machinery manufacturers for granulators.

For so many years, slow-speed granulation has been the norm with 25 rpm being the worldwide acceptable standard. A new and interesting concept by Rapid Granulator's OneCut Pro granulators which has provision for processors to adjust the rpm range from a low speed of 25 rpm to a bandwidth of 15 to 35 rpm (±40% rotor speed) to achieve optimal quality regrind.

These energy-smart machines running at 15 rpm can improve the ground brittle materials and minimize dust generation with another important factor – reduced noise. Speed can be increased at any time to 35 rpm, thus increasing the output to around 30–40% by handling larger amounts of waste materials.

The WLK 1500 shredder offered by WEIMA has the new safety features for universal operation. The modified WEIMA WAP gearbox has a safety clutch, protecting the shredder from impurities and other possible damages. Another size reduction specialist company – Piovan – offers a new cutting innovation in the U & G disposable knives system, which promises longer knife life for recyclers of plastic wastes, as well as for processors. Since these knives are very small, expensive toll steels can be used for longer life, efficiency and still cost less than conventional knives.

It is customary for a polymer processing plant to have at least one or two granulators to "recycle" the plastic wastes resulting from production. In some manufacturing processes, some of the wastes can be immediately shredded into small pieces and used immediately so no contamination is involved. This is especially so in the manufacture of "low-end" products. Other than rejects, inherent wastes come from purging of the machines before start-up and also from production setup. Downtime due to various reasons can also generate plastic wastes.

## Bibliography

[1]  Sepe, Michael. Article on "Understanding the Science of Color," 8/26/2016.
[2]  "Preventive Maintenance: 6 Elements of a Successful Program," www.mobility-work.com.
[3]  Defonseka, Chris. "Practical Guide to Flexible Polyurethane Foams," P. 83–87, 177–188, 2013.
[4]  Canadain Plastics Magazine, February 2020, www.canplastics.com.

# Chapter 7
# Lean Processing for Efficiency and Profitability

## 7.1 The Concept of Lean

Lean manufacturing or lean processing is a methodology that focuses on minimizing wastes within all phases of an operation and in this case – processing. Also known as lean production or just lean, this sociotechnical approach is based on a system that originated at the Toyota Production plants to improve efficiency through the elimination of "wastes" in excessive actions during operations which has multiple benefits ending in higher profitability. This system is so effective that many western world industrial giants have adopted this system into their operations.

Lean processing is based on a number of specific principles such as the Japanese concept of *Kaizan* meaning continuous improvement. The basic benefits of lean include reduced lead times, minimized downtime, reduced operating costs and improved product quality, to name just a few.

There are five core principles, which are *value, the value stream, flow, pull* and *improvement*. These will form the basis for implementing a lean program on a processing system, which should include all basic phases – from raw materials to transport to machine start-up, production, QC systems to packaging until warehouse storage. The ultimate result will be a good quality product of high value, and enhanced profitability for the processor.

### 7.1.1 Identifying Value

For the production of a quality product, the interaction between a customer and a processor plays a crucial role. The value is defined by the customer and this value is created by the processor or producer. The producer need to understand the value the customer places on his product or services, which helps the customer to decide how much money he is willing to pay. In order to meet a customer's optimal price, a processor or producer must strive to eliminate waste and reduce costs to make the highest profit possible. A lean system is an ideal tool to achieve this end.

### 7.1.2 Map the Value Stream

This principle involves observation, recording and analysis of the information. Starting with the required materials for a product or products, analyzing how

https://doi.org/10.1515/9783110656152-007

much is actually being used and identifying excess and methods for improvement. The value stream encompasses a product's entire lifecycle from raw material to disposal.

A producer must examine each stage separately for excess or waste (*muda* in Japanese). Anything that does not add value must be eliminated. Lean thinking recommends that supply chains also be aligned as part of this.

### 7.1.3 Creating Flow

Commence by eliminating functional barriers and identify and implement ways to improve lead time to ensure that the various stages of a process are smooth from time an order is received through delivery. Flow is critical to the elimination of waste. Lean processing relies on preventing interruptions to every phase of a production process so that all activities move in a constant stream.

### 7.1.4 Establish a Pull System

In a pull system, a new production starts only when there is a demand or if when a new order is received. Lean manufacturing uses a *pull system* instead of a *push system*. With a push system using softwares such as manufacturing resource planning, inventory needs are determined in advance and a product or products are made to meet forecasts. However, forecasts can be typically inaccurate, which can result in swings between too much and not enough inventory, which could lead to disrupted supply schedules and thus poor customer relations.

### 7.1.5 Constant Improvement

Lean manufacturing depends on the concept of continually striving for perfection, which entails targeting the root causes of excess and eliminating waste across the value stream.

According to the Toyota Production System, there are seven wastes or processes and resources that do not add value for a customer. They are as follows:
- unnecessary transportation;
- excess inventory;
- unnecessary motion of people, machinery or equipment;
- idle people, machinery and equipment – downtime;
- over production of products;
- excess production times;
- defects, which require effort and cost for corrections, and outright rejects.

Although not originally included in the Toyota System, many other lean practitioners point to an eighth waste: *waste of unused talent and ingenuity.*

### 7.1.6 Lean Versus Six Sigma

Both systems seek to eliminate waste. However, the two uses different approaches because they see the root cause for waste differently. In simple terms, lean holds that waste is caused by excessive action and processes and features that a customer believes does not add value and will not pay for, and Six Sigma holds that wastes result from process variation. Still, the two approaches are basically the same and complimentary and now have been combined as Lean Six Sigma.

## 7.2 Guidelines to Thinking Lean

Thinking lean is a way of doing a business – manufacturing or service, for example, to offer quality at customer acceptable prices in a competitive environment. A customer may pay more than your competition if you can guarantee quality at reasonable costs and satisfy the needs of a customer. In manufacturing, you must have a process to make quality products at lowest costs as quickly as possible. To do that, you need to eliminate or have processes that add value to products from a customer's perspective. This means you will strive to get rid of waste everywhere in your business.

### 7.2.1 What Are the Benefits?

In order to reap full benefits, a manufacturing company or organization must implement lean practices from the start to the final point of shipping of products. Since implementation of lean practices on an overall basis could be time-consuming, an organization may implement it on important areas only at first and then spread it gradually over a predetermined period. By thinking lean, an organization can be more efficient, and productive in any competitive environment.

When using lean practices, the basic benefits are as follows:
- productivity increases from 10% to 50%;
- quicker detection of errors and quality problems;
- reduced space requirements and shorter travel distances;
- increased ability for handling and shipping;
- reduced order processing errors;
- better customer service;
- lower material costs.

These will lead to
- significant cost savings in reduced inventory costs (30% to 80%);
- less handling of damage and lower material handling costs;
- reduced setup time (30% to 90%);
- reduced turnover and costs of attrition;
- less paperwork.

### 7.2.2 Continuous Improvement Initiatives

In order to make continuous improvements, you need to have clear objectives, detect problem areas and be able to take prompt corrective action. To meet ever-changing customer expectations, you should avail yourself of new procedures and technologies for easier operations and the desire to run a better profitable organization drives the need for continuous improvement. The process of lean never stops.

Once a lean program has been initiated on a production floor, some of the steps that can be taken to keep the momentum going are as follows:
- Lean action should be carried out by an independent team or at least a team made up of few people other than the department being "investigated." Outsiders observe the "flaws" or where there is opportunity for improvement much better than the people who are already in it.
- Alternatively, if costs are viable, an organization can engage a suitable outside consultant.
- Do the simple and less costly things first. For example, start with raw materials such as storage and reducing inventory. Perhaps a just-in-time (JIT) inventory system.
- Your floor operator's ideas/suggestions are valuable. Initiate a feedback system.
- Schedule regular meetings where possible to review and update necessary action.
- Feedback from your customers is very important for maintaining product quality.
- Make continuous improvement a part of your business culture.

Despite the popular conception that lean production is a complex and costly process, it is not. Once you get the cooperation of your employees and the need for lean practices, especially in highly competitive market environments, everybody will find it as both exciting and challenging work which will result in producing quality products at lesser costs, which of course, means higher profits. It is up to the management of an organization to offer incentives to employees.

### 7.2.3 Push–Pull System

In production systems, basically either production is made against customer orders or made for stock against forecasts or even both may apply. The conventional "push" system where large volumes of products are made, perhaps against a free-flow market, will keep people and machines occupied but producing in this fashion could create waste, if for example, market customer satisfaction, lower inventory and lower costs, especially if customer product designs change frequently.

During production planning it is important to take into consideration the possible complexity and product diversity spanning over a long period. The pull system works on the basis that there is no disruption in a production process from beginning to end. If your product is relatively simple, you can make the necessary adjustments to improve and set higher standards. If parts are made for an assembly line, there could be sectional delays which will lead to production delays. If large volume assemblies are involved, defective parts could be made for long periods before the defects are detected. This is where lean practices will help.

Implement an action plan with targets that can be achieved gradually over a period of time. Constantly monitor and make more adjustments where necessary. The best chance for success lies in first observing and then taking corrective action as you go along.

### 7.2.4 Kanban System

It is a method for maintaining an orderly flow of material to ensure continuous production without material shortages. Popular methods are card systems covering the various stages of need and effective ordering levels. Naturally, an organization would like minimum inventory levels for reduced costs and may opt for JIT Systems. However, depending on the availability of materials and advantages in large bulk purchases, JIT systems may not be viable.

## Bibliography

[1]   Business Development Corporation, Canada article on "A Quick Guide to thinking Lean", BDC.ca-2019.

# Chapter 8
# Processing Systems for Polymers

## 8.1 Some Common Systems

This section covers some of the common processing systems used to convert polymer resins into plastic products. These will vary from the small hand-operated machines for injection molded and blow molded products to the more sophisticated manual, semiautomatic and automatic molding, foaming, coating machines and so on.

The choice of a process for a processor will depend on whether a product is to be made with a *thermoforming* or *thermosetting* polymer resin. Some thermoforming polymers commonly used are polyethylenes (PEs), polypropylenes (PPs), polystyrenes, polyvinyl chlorides (PVCs) and others, while some thermosets are melamine formaldehyde, urea formaldehyde, polyurethanes (PURs), silicones and others.

The size of a product to be made also has some bearing on a process selection. For example, large products like industrial bins, children's playground items, industrial utility carts, laundry trolleys and even much larger volume products are produced using a rotational molding process (presented in detail later). Once a customer has visited and examined a production plant and has placed an order, the normal standard procedure is to set a production part approval process procedure in motion, where a processor will make a sample/samples and submit it to a customer for approval before a production run starts. Here, two of the basic important systems on a production floor will be quality control procedures before parts are shipped to customer to prevent rejection, and returns and preventive maintenance of all machinery and equipment to minimize downtime and enhance efficiency and productivity for higher profitability.

## 8.2 Common Processing Systems for Thermoforming Resins

By far, the greater percentage of polymer resins processed into plastic products probably are thermoforming resins, although very large volumes of PUR foams are also being made. Presented here are some established processes for these types of resins.

### 8.2.1 Extrusion

Extrusion systems have come a long way from the original to the modern precision extrusion systems and are designed to meet challenges like high outputs, quality, cost-effectiveness, easy setup and operation through a combination of monitoring, data collection, self-correcting automatic process controls and so on. These modern

https://doi.org/10.1515/9783110656152-008

systems use well-advanced and constantly improving technology and are versatile enough to process the ever-increasing range of polymer resins and polymer blends.

Extrusion systems are basically made up of an extruder with a hopper, a barrel with three or four heating zones containing a rotating screw, a die head, orientation unit, cooling arrangement, take-up or conveyor system and, finally, a winding unit for some. This line can also have other ancillary equipment like blow rings, cutters, formers, printers, sealers or other equipment depending on the process. Extruders can be single screw, twin screw or have multiple screws for very large extrusions or for streaming more than one material. Screw diameters can range from 12 to 300 mm. Common single screw will have diameters from 25 to 200 mm and may have grooved barrels, vented barrels, tiltable and tandem specialty screws and in addition to the most common horizontal operations, some extruders may be vertical for special applications. The zonal heating bands may require periodic replacement, and the screw and barrel also undergo wear and tear from constant usage. To overcome this, modern extruder manufacturers use protective coatings for both.

Extruders are generally classified in relation to the screw – $L/D$ (length-to-diameter ratio) and may range from 5:1 to 40:1 with an average of 24:1 to 30:1 being ideal for standard processing of polymers. To ensure product quality, the selection and use of the right screw in relation to the polymer is important and a processing plant may have to change screws for different materials. However, universal screws are available, which can process most polymers. Screw speeds may vary from 25 to 250 rpm or other.

Irrespective of screw size or other, all extruders have a same operating principle. A hopper mounted at one end is filled with the polymer material and will have direct access to the barrel via an open/close arrangement. Some hopper may have a vibrating system to ensure smooth gravity-feed into the barrel but all hoppers will have a "window" where an operator can see the level of material. The material is first fed into the "feed" zone of the barrel containing a rotating spiral screw and the hot melt is slowly carried forward through the other zones, which turns it into a homogenous free-flowing mass and reaches a mesh pack with breaker plate which prevents any foreign matter from getting through to the die and the constant rotational force will force the molten material through the orifice of a die in the die head. In the case of extruded pipes, for example, the extrudate (extruded profile) will go through a sizing/calibration arrangement and the emerging pipe will be cooled immediately and cut to desired lengths.

In the case of any wire (cables) or other is to be coated, the wire will be fed at right angles to the flow of the molten polymer mass using a cross-head die and the output will be very fast and at least a two-station winder will have to be used.

Modern-day extruder manufacturers employ two basic approaches to ensure quality and productivity. The first is to ensure product uniformity by using higher precision motor drive controllers and better temperature controllers to ensure constant uniform flow. The second is through the use of closed-loop controls with autofeedback and autocorrections to preset parameters. These systems are far superior to the earlier electromechanical and subsequent open-loop controls. Figure 8.1 shows basics of a single-screw extruder.

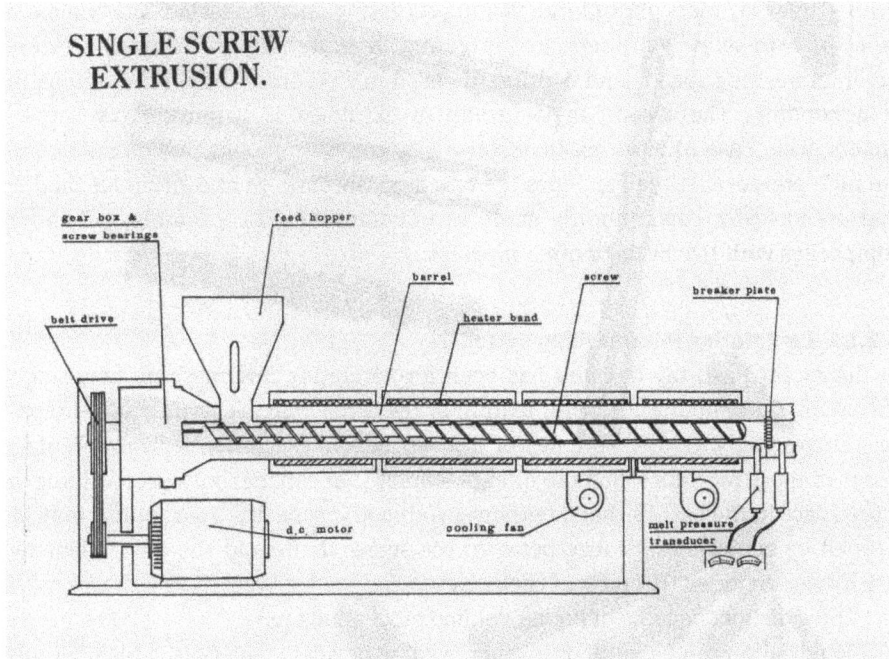

**Figure 8.1:** Basics of a single-screw extruder.

A hard fact about extrusion processing is that feed-screws and barrels deteriorate over constant use and need to be repaired or replaced in order to maintain quality and production levels. By understanding the warning signs and corrective action in advance, the damage can be minimized. The first thing in fighting back against barrel and screw wear is to understand how and why they happen. Barrel and screw wear occurs between the top of the screw flight and the inside diameter of the barrel. As the amount of wear increases, the screw outside diameter (OD) gets smaller and the barrel inner diameter (ID) gets larger, creating more clearance between the screw flights and the barrel. Allowing the extruder to work in this condition means that the increased clearance allows the hot plastic melt to pass over the flights of the screw, instead of pushing it forward.

This can cause uneven feed to the die, and thus micro- or even larger variations in the extrudate. In difficult to process resins, this effect can get magnified. Three conditions define these and can be attributed to *adhesive wear, abrasive wear* and *corrosive wear*, and can be overcome/minimized by proper treatments for both barrel and screw-like suitable protective coatings.

### 8.2.1.1 Some Types of Commercial Extruders
Twin-screw extruders play a large role in processing polymer resins. The two main types of twin-screw extruders are corotating intermeshing and counter-rotating nonintermeshing speeds and is primarily used in PVC processing, for example, in compounding. The second major group of extruders is twin-rotor continuous mixers, which use rotor rather than screw to accomplish mixing and do not generate high pressure. These machines are based on the concept of continuous mixing and are ideal for compounding mixtures of composites, for example, polymeric composites with rice hulls or other biomass.

### 8.2.1.2 Co-Rotating Intermeshing Extruders
In the recent past, the demand has been for corotating intermeshing twin-screw extruders. One important overall feature of these machines is their ability to process increasingly greater volumes of a given size of machine. Several manufacturers of these machines claim to offer machines that can maximize free volume or more space in the barrels that determines volumetric capacity. Free volume may be defined by centerline distance between the screw shafts and the outer diameter (OD)/inner diameter (ID) ratio of each screw.

Three distinct benefits of higher volume processing are:
–  Fluidization can be minimized when feeding powdery materials, which permits higher throughput rates.
–  Residence time can be maximized, allowing higher throughput of highly divergent viscosities.
–  Vacuum entrainment can be minimized, allowing higher volatile levels to be drawn from the process.

According to some designs, some screws have two channel depths at the same screw diameter, whereas in some, the screw flights are deeper and the root of the screw is reduced. According to manufacturers, the larger free volume, coupled with high torque capability, results in higher throughput. Another important aspect of the free volume issue is the ability to transmit power through the shaft to the material. The ratios between free volumes and power will vary with different designs and different manufacturers.

Another related trend in corotating intermeshing twin-screws is their ability to run at increasingly higher speeds. These speeds are only limited by whether it is possible to maintain quality at these high speeds. Other limiting factors may be wear related as with abrasive fillers. Composites material processing, for example, especially with high reinforcing levels with rice hulls may pose a problem, since rice hulls contain around 20% silica, which will generate extra heat. However, extruder manufacturers have overcome this problem with specially designed screws. For nonabrasive mixtures, high speeds can be achieved easily. Shear stress can be often minimized by screw design, so that even temperature-sensitive materials like polycarbonates (PCs) can benefit from higher outputs.

### 8.2.1.3 Counter-Rotating Nonintermeshing Extruders

Manufacturers of these machines provide several advantages to feeding, mixing and venting with some machines that can be converted to all types of operating modes. One advantage is maximum free volume, which is the result of two design factors inherent in nonintermeshing screws such as: (1) greater centerline distance can be achieved between screw shafts and (2) flight occlusion. A third point relates to flexibility. Because the screws do not intermesh, it is possible to vary the root diameter to provide much-needed shear energy for a given application. Some school of thought estimates that these nonintermeshing machines can give 30% more output than intermeshing machines of the same screw diameter. The nonintermeshing screw configurations have implications for power generation. This extra available energy for the same screw diameter enables its machines to handle greater speed, volume and length-to-diameter ratio (L/D).

Counter-rotating nonintermeshing machines are well suited to enhance mixing and also provide a high degree of flexibility and control of the shear energy imparted. One advantage of the low shear in the apex is that it reportedly permits relatively high screw speeds in compounding shear-sensitive materials. If more shear is needed, compounding sections can be added. Staggered screw flights provide cross-channel and radial back mixing. These machines will also have specially designed feed zones and wider venting directly above the apex.

### 8.2.1.4 Maintenance of Extruders

As a general experience, all extrusion processing undergo deterioration of the barrels and feedscrews. If abrasive materials are processed, this deterioration will naturally be faster. However, by understanding the causes for these and learning to see the warning signs, a processor can minimize the damage.

The first step in controlling wear of the barrels and screws is to understand why they happen. Barrel and screw wear occurs between the top of the screw flight and the inside diameter of the barrel. As the amount of wear increases, the screw OD gets smaller and the barrel ID gets bigger, creating more space between the screw flights and the barrel. Operating an extruder under these conditions means that the increased space allows the plastic material in the barrel to pass over the screw flights, remain more or less unmixed instead of moving forward. There are different causes for this but according to statistics, the main reason is misaligned machine components.

The second reason is the use of difficult materials to process or the use of abrasive materials. According to reports, crystalline engineered polymers, fractional melt polymers, abrasive fillers, potentially corrosive polymers and some others are conducive for barrel and screw wear. The third reason for wear can be foreign matter, especially metal pieces in the material used coming through the hopper of the extruder and entering the barrel.

There are three types of wear – *adhesive, abrasive* and *corrosive*. Adhesive wear is caused by metal-to-metal contact with two metals rubbing together hard enough to remove material from the less wear-resistant surface. Abrasive wear is due to microchipping and occurs when foreign or abrasive particles in the resin comes into contact with the barrel or screw. The "scouring effect" of these hard particles wears away the metal, most often in the transition zone of the screw. Some of the abrasive particles in the plastic resins can be reinforcing agents such as glass fibers, glass spheres, calcium carbonate, ceramics or powdered metals. A good example is when processing polymers with rice hulls powder or flour for the manufacture of composite resins. The content of silica in the rice hulls will act as an abrasive material. Wear could also occur when processing nonreinforced polymers if too much energy, to melt the polymer, is generated by excess shear. Cold resin pellets moving into the transition section of the barrel are compressed and sheared causing a "*scrubbing effect*" which causes wear. Meanwhile, corrosive wear occurs by chemical reactions inside the processing unit and is characterized by "*pitting*" and usually occurs in the last few flights of the transition and metering zones of the screw. These "pits" can burn or degrade the melt, causing black spots in the extrudate.

According to the experts, when extruder wear occurs there are early warnings which can be easily recognized by an experienced operator or a floor engineer. Some of them are the need for increased screw speed to maintain normal output, above normal melt temperatures, material back up in the feed zone, variations in output and poor product quality. Also, primary streaks and dark specks are also indicative of extruder wear. Probably the most obvious sign observed by an extruder operator is the increase in the scrap rate. Generally, extruders work very smoothly and if an operator hears any unusual sounds like grinding, it should be reported to a supervisor.

### 8.2.1.5 Some New Technologies in Extrusion

Extrusion has always been both an exciting and challenging field of processing for plastics, and extruder designers and manufacturers have always been working hard to come with new technologies to boost output, reduce energy usage, design more efficient dies, barrels and screws to improve product quality with costs also in mind. Moreover, over the years, extruder manufacturers have had to deal with new polymers coming on the market in addition to the standard ones.

Here are some new technologies offered. battenfeld–cincinnati offers a new design, single screw extruder – solEX NG, which has a completely redesigned screw feed zone with a significant low-pressure profile. This enables high specific outputs, fast process start-ups at low screw torque and no conveying instabilities, even at high back pressures (7,200 psi). This design also allows reduction of melt temperature of up to 20 °F. This means that the cooling length of the extrudate can be shortened as the cooling baths for the die-emerging extrudate need to remove less heat. It is also possible to increase the line speed, thus, resulting in output increases of around 20%.

Another extruder manufacturer, Leistritz Extrusions offers a new-in-line compounding of polymers with additives and active fillers. This new extruder – ZSE-3D – is a twin-screw extrusion system designed for production of 3D filaments from a corotating or counter-rotating screw system. This machine is ideal, according to the manufacturer, for in-line compounding of polymers with additives and active fillers. Formulations can be modified to include water-soluble and high-temperature engineered polymers.

New from Milacron Extrusion Technologies is the SV350 designed to be a robust, flexible extruder for processing profiles, tubing, sheet, fiber, wire and cable. The screw range for these single-screw machines is from 2.0 to 4.5 in with an *L/D* ratio of 24:1. The main advantage of this extrusion system is its versatility to handle a multitude of applications and materials. Their range of supplies includes new and rebuilt extruders, barrels, screws, pipe heads, dies and down-stream equipment.

Here, the author provides the manufacture of three plastic products to demonstrate the process of extrusion for the benefit of the readers.

**Plastic Rattan:** These are needed in very large quantities for chairs, loungers, furniture and so on in various cross-sections and in white and other colors. The process in brief is a – profile extrusion, immersion in a hot-water bath for orientation (stretching) and cooling. First, the producer will decide on the cross section of the profile and size. General stretching ratios are 1:3 or 1:5, which will indicate the required starting profile section size. A small 25 mm extruder can be used with a suitable die profile which is generally semicircular.

For production of white standard rattan for weaving on chairs, loungers and others, a monofilament grade of high-density polyethylene (HDPE) can be used, while for colored and slightly bigger rattan, PVC can be used with either self-colored or colored with masterbatch. The latter will not require stretching and can be directly extruded onto a conveyor cooled and rolled up. In the case of the white monofilament being extruded, it will go through a water tank at 100 °C and gets stretched between squeeze rollers at the point of entry and exit, where adjustments are made to stretch the hot monofilament to a predetermined ratio. This extrudate will then go through a cooling system.

If a "black" colored base is preferred to counter UV action, an additional extruder can be used in conjunction, feeding the same die and in a coextrusion process. The stretching process must be done properly as otherwise "splitting" will occur.

**Plastic conduit pipe:** It is used in electrical work and for other industrial purposes and can be made using a standard extruder. Depending on the diameter of the pipe, generally, a 40 mm extruder would suffice with a die arrangement with a round mandrel around which the hot plastic melt would flow forming a tube. The diameter of the mandrel core will be the same as the inner diameter of the pipe. The die will also have an outer ring and the gap between the mandrel and the ring should be slightly larger than the outer diameter of the pipe.

The color of the plastic resin being extruded can be in accordance with any color codes desired such as white, gray and black and is fed into the heated barrel of the extruder via a vertical hopper. This material will pass through the mixing zone, the plasticizing zone and finally the metering zone before it reaches a mesh pack screen to prevent any foreign matter passing through. As the hollow extrudate emerges from the die, it will enter a "sizing" arrangement which will determine the final dimensions of the pipe.

As the pipe travels forward, it will still be warm and a polished and engraved rotary disk dipping in a small ink tank touching the bottom surface of the pipe will print a brand name, specs and other information. A cutter mounted on the conveyor will cut the continuous pipe into desired lengths.

**PVC electric cables:** The manufacture of electrical cables can take the form of a simple extrusion operation to produce cables for the domestic/residential markets to more sophisticated machinery and equipment for industrial applications, then again specialized manufacturing for very advanced applications where strict quality, tolerances, durability and properties are required more than for others. However, since we are dealing with the conductance of electricity, strict adherence to whatever international standard they are being made to must be achieved. Here, the author presents the basic manufacturing operation for cables suitable for 5 , 15 and 60 amp use. This will demonstrate the basic principle of making electrical cables for the benefit of the reader.

Depending on the selection of PVC coating lines (continuous or two stage), the design of the production factory will be more on length than width as the coating lines are long. Basic factory floor design should include space for:
- store room for wire,
- store room for polymer resin bags or totes,
- large area for wire preparation,
- extrusion area for two extrusion lines,
- finished cables testing room,
- packing area,
- area for QC inspection,
- finished goods store room,

- shipping area,
- offices and others.

As an example: Production of cable sizes – 1/.044, 3/.036, 7/.029 (earth cable) 7/.064 (twin) 14/0.0023. The first number indicates the number of wires, while the second indicates the gauge or the diameter of the copper conducting wire. Both copper and aluminum can be used with copper being the preferred one and less expensive and has a better conducting value than aluminum. To manufacture these PVC cables, a manufacturer will chooses an internationally accepted standard like British Standard Specs (BSS), American Standards for Testing Materials (ASTM), DIN (German standards), JIS (Japanese) and so on.

The basic extrusion lines and equipment needed for this operation would be as follows:
- 40 mm extruder (A) with cross-head, autotension wire feeder system, take-up unit + winder with autotraverse;
- 65 mm extruder (B) with cross-head, coated cable feeder, electronic centering device, engraved steel roller (print brand name), take-up unit, high-voltage tester, + winder with autotraverse;
- 1 No. bunching machine for small gauge multiple wires;
- 1 No. stranding machine for larger gauge multiple wires;
- large wooden drums for winding coated cables;
- suitable wooden drums for winding coated cables;
- suitable packing equipment.

Whether the production is for a single or multiple core wires, the principle of manufacture is more or less the same. The production is carried out in two stages. The first one will give an initial coat-extruder A, the thickness and color depending on the standards used. It is important to see that the core wire/wires are centered to the coating. The processing speeds are very fast and if the core is not centered properly, many rejects will result. These first insulation-coated wires are taken up in large wooden drums with a traverse in use to ensure a smooth, untangled winding.

These drums are then brought to extruder B and fed through the cross-head for the second and final coating. Again, the thickness and color will depend on the standard used. The small printing roller placed under the coated cable will print a logo, brand name or others as desired. Here, the tension on the take-up equipment must be adjusted so as not to "stretch" the copper wire. This is to ensure there is no change in diameter size and also to prevent breaking of the conductor. The coated cable now goes through a high-voltage tester to ensure good insulation. If there is a defect, an alarm will go off and automatically the surface of the defective area will be marked which can be cut-off at the packing stage. As in the first stage, the cable will now be fed into a large wooden drum with a traverse.

Standard packing lengths may be 50 m or 100 m but if the production has too many coating or other defects, there will be "short lengths," making it difficult to market. Constant testing by the testing lab will ensure good quality products.

### 8.2.1.6 Recommended Shutdown Procedures

The key to a successful shutdown is to purge the current resin in the barrel operating temperature with a more stable resin introduced at a colder temperature. This can be achieved by reducing the operating speed at the end of a run and lowering the temperature to a safe level. The system can now be purged with material of good thermal stability. A base resin blended with an antioxidant master batch will work well for this procedure.

- Reduce the screen pack mesh if necessary.
- Add the purge blend into the hopper.
- Reduce operating speed by 75–100%.
- Lower barrel temperature to 325 °F (general) – 375 °F for PP.
- Back-out valuable depth thermocouples.
- Add shutdown blend.
- Shutdown at around 25–40% normal output.
- Drop barrel temperature further to 280–350 °F for PP.
- Watch head pressure as barrel temperatures drop.
- When the purging material appears out of the die, be ready for shutdown.
- Leave a small amount of material over the die to prevent oxygen from getting into the die lip land area.

### 8.2.1.7 Recommended Restart Procedures

To restart, bring the line to start-up conditions at a lower temperature than normal operating temperature and heat-soak the line to ensure the material in the barrel is molten before restart. The reason for this is twofold. One is to reduce exposure of the stagnant polymer to a higher temperature. The second is to maintain a higher polymer viscosity at the lower temperature. The higher viscosity creates scrubbing forces against the flow surfaces, assisting with the purge. A shutdown with a thermally stable material will also allow more time and flexibility on the restart.

The following is a recommended procedure for an extrusion-blown film line processing linear low-density polyethylene (LLDPE). This example can be adapted in general on the basis that the heat up of a large extrusion system will be governed by the heat-up rate of the die system.

- Set die temperature to 250 °F.
- Soak die for 1 h.
- Increase die temperature to 300 °F.
- Turn on adapters to 250 °F.

- Soak 15 min once set point is reached.
- Increase die to 350 °F.
- Increase adapters and screen changer to 325 °F.
- Turn on the extruder 250 °F.
- Soak for 15 min.
- Increase the die to 375 °F.
- Increase the die changer to 375 °F.
- Increase extruder to 300 °F.
- Set extruder to 325–340-350-350-350 °F zonal start temperatures.
- Heat soak for 15 min.
- Start extruder at a very low speed of around 10 rpm.
- After the material is clear, change screen pack to the production mesh.

Note: The above is presented as guidelines only to show the general procedures related to standard extrusion systems. Naturally, these parameters will vary with different systems and a processor can be guided by the specific information provided by machinery suppliers. The main criteria will depend on the material to be processed, the screw type and so on.

**Polymers in flooring:** Polymer floorings, especially, vinyl flooring has been in use for decades. At the early stages, vinyl flooring was individually molded to various thicknesses and self-colored. Later embossed surface patterns were available. As the technology developed over the years and the demand increased, the manufacture of vinyl flooring became an extrusion process, where wide extrudates were possible, cooled and then cut to desired sizes.

PVC being easy to process, made fast extrusion rates possible. Modern-day technologies include biomass-filled flooring and also pleasing laminated veneers for aesthetic efforts to imitate natural wood grain finishes for flooring.

## 8.2.2 Injection Molding

Injection molding is probably the most popular processing system for polymer resins, estimated to be around 70% of all processing. Injection molding will range from the humble hand-operated molding machines to manual, semiautomatic and fully automatic machines. Over the years, the injection molding process has developed from simple manually operated plunger-type machines to the current very sophisticated screw injection machines. Moreover, with advances in technology, current machines are as hydraulic, electric, hybrid machines and so on. General classification of injection molding machines depend on the shot size and clamping force.

Plastic injection molding is a process commonly used for making solid plastic parts from small trinkets to cell phones cases, water bottle containers, consumer items, to even larger parts for the automobile industry, structural foam and solid plastic pallets and many others. It is a quick process for single or large volume productions with multicavity molds. Basically, the process is the injection of a molten mass of polymer resin into a closed mold in two halves, with cooling and ejector arrangement.

An injection machine is made up of four primary main components – a feed hopper, a screw, a heated barrel and a mold mounted on two platens. The plastic resins normally used are as pellets or powders and gravity fed into the barrel containing the screw. The heated barrel melts the polymer aided by the friction of the rotating screw, turning it into a homogenous molten mass. This mass is carried forward by the rotating screw and then injected through a nozzle at the end of the screw into the opening of a mold. The material is "held" in the closed mold during a specific preset time, auto-cooled and as one half of the mold opens the part/parts are ejected.

Injection molding machines are generally rated by their clamping force (tonnage) and also their shot capacity, meaning the maximum volume/weight of molten plastic that can be injected through the nozzle. This will be around 70–80% of the rated capacity and the actual injection shot will be determined by a preset "cushion" calculated to inject sufficient material for either a single or multicavity production process.

Injection molding machine's sizes, for example, can run from 5 tons to 6,000 tons of clamping force. The higher the tonnage, the larger the machine. In injection molding, another important factor is the viscosity of the mass being injected which will vary with different polymers. The melt flow index (MFI) is a measure of the melt of a thermoplastic polymer indicating the ease of flow. The greater the MFI, the higher the required tonnage. For example, if a multicavity mold has four cavities, with each part 5 × 5 × 0.2 sq in. First, calculate the projected area of the part. With this mold, the calculation would be 5 × 5 = 25 sq in x 4 = 100 sq in of projected area. This factor is needed because it is the principal variable that affects clamp tonnage. If a 10 MFI PP is used, a minimum of 2.5 tons per square inch of projected area would be practical. Therefore, for this mold production a machine with at least 250 tons would be needed.

Some of the defects of parts in injection molding are splay, warpage, short, uneven color, flash and some others. With constant advances in technology, modern machines are very sophisticated with ease of operation, faster productions and cost savings. Emerging technologies are allowing molders to use 60:40 or even higher filled composite polymer resins on normal standard machines with different operating parameters.

### 8.2.2.1 Structural Web Injection Molding

Structural web molding technology is a variation of low-pressure injection molding (LPIM) that was developed by Milacron as a molding process on their LPIM machines.

It is a process well-suited for making large solid parts with good part surface aesthetics low molded stresses. This process can also create parts with a dual-wall hollow sections, thus reducing the overall weight of a large part by 10–30% within the same geometric pattern.

With Milacron's LPIM machines being multinozzle injecting machines, structural web molding with options are possible. This process is a multipoint gas-assisted process that introduces nitrogen into a molten polymer mass after injection into a mold. Unlike standard gas-assisted processes, these machines do not need gas pins in the mold, as it is introduced through one or more structural web LPIM nozzles. Figure 8.2 shows a three-nozzle arrangement:

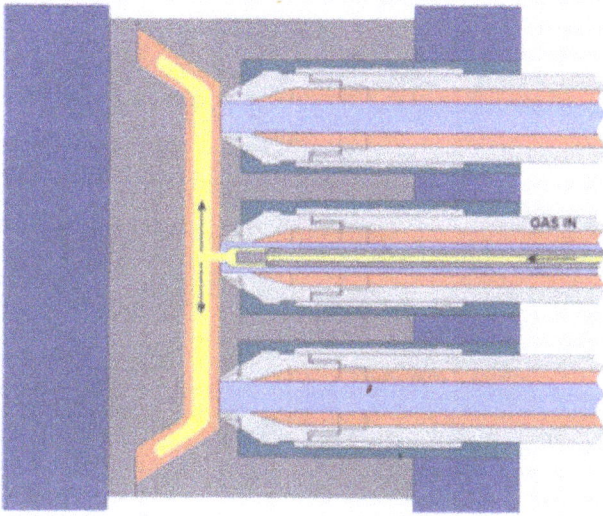

**Figure 8.2:** Three-nozzle arrangement reproduced with permission from Milacron.

In a structural web process, the nozzles open and partially fill a mold cavity. Then, gas is introduced into the molten mass in the cavity and pushes the material against the cavity walls, filling out the full cavity. Nitrogen gas is allowed to vent out of the part and the part is ejected after sufficient cooling.

### 8.2.2.2 Benefits of Structural Web Processes

This process allows for molding parts with thick or varying wall thicknesses such as a thicker rib intersection with the main structure of the part, allowing for a stronger part but without sink marks. Also, since this process is for solid polymer parts, there will be no swirl marks on the surface unlike in structural foam processes.

Parts can be easily molded in color and desired textures, and will not require post-molding finishing, for example, like decoration.

LPIM machines are generally large volume shot size machines – from around 90 kg and up – allowing for ultralarge molding which is not possible in standard injection molding methods. Two of the most popular polymer resins used are HDPE and PP, but engineering resins can also be processed successfully on these machines. Typical products are utility carts, large rubbish bins, mop buckets, material handling pallets, construction drainage chambers, cargo bins and medical equipment housing cabinets, to name a few as shown in Figure 8.3.

**Figure 8.3:** Products reproduced with permission from Milacron.

### 8.2.2.3 Structural Foam Injection Molding

Structural foam injection molding is a low-pressure injection process, where an inert gas is introduced into the hot polymer mass for the purpose of reducing density, and hence reduce the weight of the finished part. These parts will have cellular cores covered by rigid, integral skins. The gas is introduced into a short shot (less than actual weight of part) and then injected into a mold through nozzles. Injection pressure and expansion of the polymer/gas mass will then fill the entire cavity of the mold. A skin will form when the melt comes in contact with the cold surface of the mold, while the expanding polymer/gas mass will form a cellular core.

Some of the advantages are:
- Part weight reduction from 10% to 30%
- Density reduction – savings in resin costs
- Use of nitrogen gas or carbon dioxide gas costs less than use of chemical blowing agents

- Large parts molding with low-clamp force requirements
- Mold cavity pressure typically – 200–600 psi (14–41 bar)
- Lower energy costs versus other injection processes
- Lower cost aluminum molds versus high-pressure IM machines
- Faster cycles – better heat transfer of aluminum
- Thick wall parts – 3–12 mm
- Complex parts without sink marks
- Higher impact strength than thinner wall IM parts
- Parts can be sawn, screwed, nailed or stapled like wood

### 8.2.2.4 Hand-Operated Injection Molding

This is an ideal operation for an entrepreneurial venture, which is essentially a good method for making small parts of low volumes. For an entrepreneur with some basic engineering skills, these machines could be designed by himself and fabricated on a production floor itself. They could be operated manually, electrically or hydraulically but depending on the design, the shot capacity would be limited. However, it is recommended that the tooling (mold in two halves) should be made by a professional source.

Depending on the specifications and the aesthetic values of the product to be made, the raw material could be plastic wastes or virgin material or a combination of both. Figure 8.4 shows two types of machines.

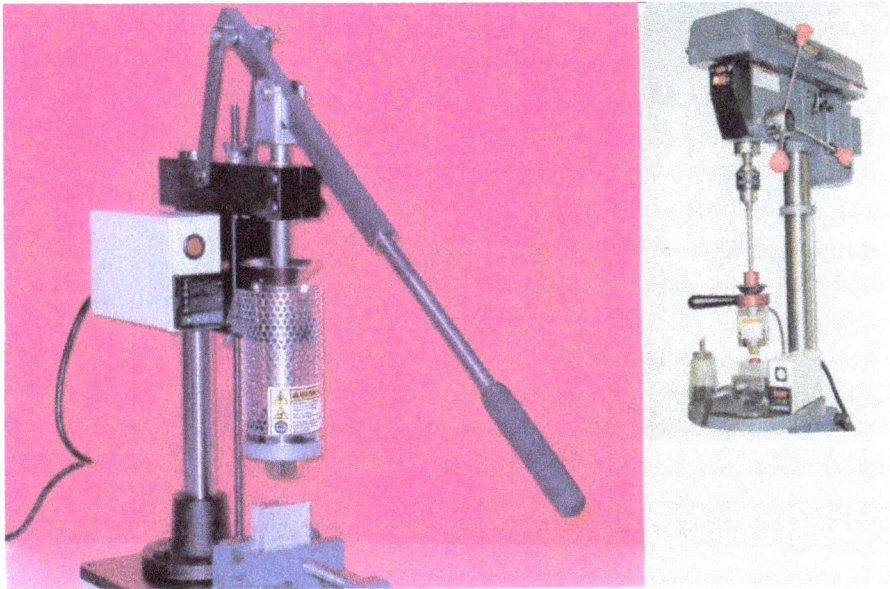

**Figure 8.4:** Hand-operated IM machines used by author to make plastic parts.

### 8.2.3 Blow Molding

Blow molding is a manufacturing process that is used to make plastic hollow parts. The standard polymer resins used in a blow molding process are thermoplastics in the form of small pellets or granules. Basically, this process involves the melting of a polymer mass to form a tube called a *parison* which is clamped between two halves of a mold and a jet of air "blows" the parison against the wall cavity of the mold taking its shape and the product is ejected as the two halves of the mold open. Typical pressures are 25 to 150 psi, far less than for injection molding.

Products made from blow molding can range from small to large such as bottles, containers, bins and so on in a variety of shapes and sizes. Small products may include bottles for water, liquid soap, shampoos, motor oil and milk packs, while larger containers include plastic drums, tubs and storage tanks. Figure 8.5 shows a blow molding sequence for bottles:

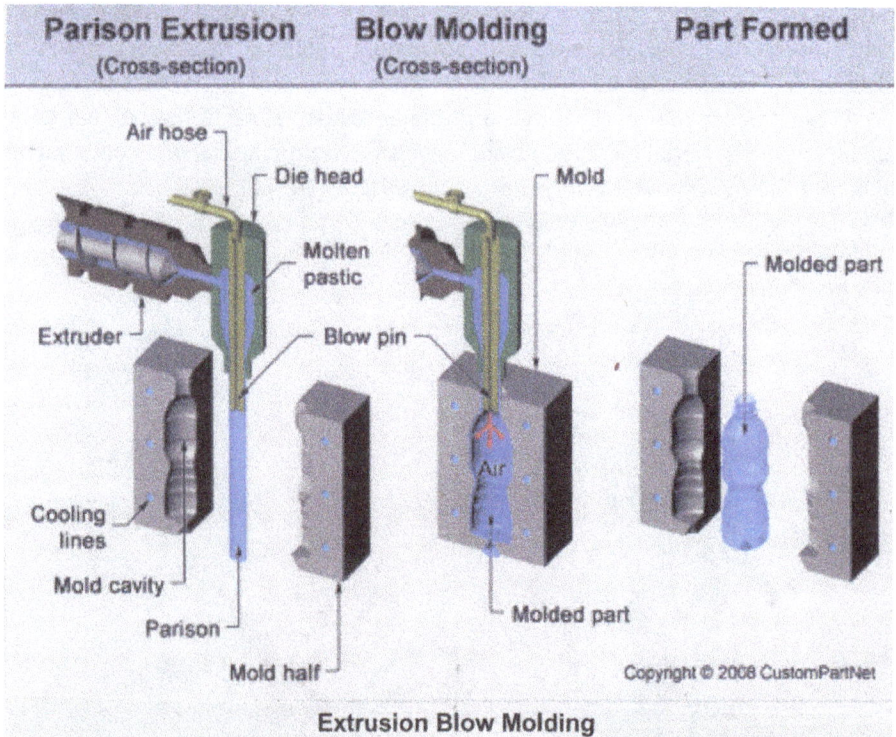

Figure 8.5: Blow molding sequence for bottles (reproduced with permission from Custom-Pak).

Some of the common polymer resins used are:
- Low Density Polyethylene (LDPE)
- High Density Polyethylene (HDPE)
- Polyethylene terephthalate (PET)
- Polypropylene (PP)
- Polyvinyl Chloride (PVC)
- Polycarbonate (PC)

Blow molding system can vary depending on the products being made, and includes automatic production cycles also with a multimold system in rotary movement or otherwise. There are different methods used to form a parison such as:

- *Extrusion blow molding:* A standard extruder is used to force the molten polymer mass through a die head which forms the parison. In the case of blown film, the thickness and the diameter of the parison determine the final parameters of the film like thickness (gauge) and width. In a vertical operation, the molten polymer is collected by an operator to form a "bubble" and fed into a slowly rotating "nip rollers." In the middle of the ring die, an opening blows a jet of air inflating the polymer bubble to predetermined ratios until the correct desired film thickness or gauge is achieved. In modern-day advanced machinery, these parameters can be preset to minimize error.

  The bubble forms a flay film tube as it goes through the nip rollers with air cooling before and is then taken up in a winding unit. Fully auto machines, for example, making plastic bags will have a printing station and finally a bag making station.

  In the case of multilayered film productions, a parison will have several layers of materials from different extruders all feeding the same ring die. These films can be processed either vertically or downwards but the former is the more popular process.

- *Injection Blow Molding:* The molten plastic mass is injection molded around a core inside a parison mold to form a hollow parison. When the parison mold opens, both the parison and the core are transferred to the blow mold and clamped securely. The core then opens and allows pressurized air to inflate the parison. This is the least commonly used process because of the lower production rate but is capable of forming more complicated parts with higher accuracy. For example, small intricate hollow products.

- *Stretch blow molding:* The parison is formed the same way as injection blow molding. However, once transferred to the blow mold, it is heated and stretched downwards by the core before being inflated. This stretching provides greater strength to the plastic walls. Stretch blow molding is typically used to create/ produce products that must withstand some internal pressure or be very durable, such as soda bottles.

Some comparative values of polymer resins as a guide are provided in Table 8.1.

**Table 8.1:** Comparative values of polymer resins as a guide (adapted from technical values as supplied by Custom – Pak Inc.).

| Resin | Density (g/cc) | Low temp. (°F) | High temp. (°F) | Flex modulus (MPaa) | Shore hardness (D) |
|---|---|---|---|---|---|
| HDPE | 0.95 | −75 | 160 | 1,170 | 65 |
| LDPE | 0.92 | −80 | 140 | 275 | 55 |
| PP | 0.90 | 0 | 170 | 1,030 | 75 |
| PVC | 1.30 | −20 | 175 | 2,300 | 50 |
| PET | 1.30 | −40 | 160 | 3,400 | 80 |
| TPE | 0.95 | −18 | 185 | 2,400 | 50 |
| ABS | 1.20 | −40 | 190 | 2,680 | 85 |
| PPO | 1.10 | −40 | 250 | 2,550 | 83 |
| PA | 1.13 | −40 | 336 | 2,900 | 77 |
| PC | 1.20 | −40 | 290 | 2,350 | 82 |
| Polyester | 1.20 | −40 | 160 | 2,350 | 82 |
| Acrylic-styrene | 1.00 | −30 | 200 | 2,206 | 85 |

Resin producers and compounders constantly come up with new and modified polymer resins with enhanced properties. For example, thermoplastic polyester elastomers (TPC-ET), a thermoplastic elastomer of copolyester is replacing traditional thermoplastic elastomers (TPE's) in elevated temperature conditions. New thermoplastic urethane (TPU) elastomers resist oils, wear and tear better than traditional TPE. For processors, it is always an advantage to work closely with polymer resin suppliers to take advantage of the best-suited resin for their intended products. The above data is to be used only as a guideline and may vary with different compounders/suppliers.

## 8.2.4 Compression Molding

Compression molding is a method of molding plastic resins, both thermoplastic and thermosets in a two-part mold system where a preheated resin is introduced into a preheated open mold and cured in a closed mold. A basic molding cycle would be heating, cooling, dwell time and then manual or autoejection. Thermosets take precedent over thermoplastic resins in compression molding and are generally popular with insert molding and very intricate patterns.

Compression molding encompasses different molding techniques such as the basic compression molding, the transfer molding process, resin transfer molding process and compression transfer molding process. These different methods provide different capabilities to fabricate/mold products to meet performance requirements using different materials.

Table 8.2 shows some of the thermoset resins used in industry and their end applications.

**Table 8.2:** Thermoset resins used in industry and their end applications (modified and adapted from Science Direct Topics).

| Material | Performance | Application |
|---|---|---|
| Phenol–formaldehyde | | |
| General purpose | Durable, low cost | Small housings |
| Electrical grade | High dielectric strength | Circuit breakers, plug tops |
| Heat resistant | Low heat distortion | Stove knobs, handles |
| Impact resistant | Strength | Appliance handles |
| Urea–formaldehyde | Stable colors | Kitchen appliance |
| Melamine–formaldehyde | Hard surface | Plastics dinnerware |
| Alkyd | Arc resistant | Electrical switchgear |
| Polyester | Arc resistant | Electrical switchgear |
| Diallyl phthalate | High dielectric strength | Multiple connectors |
| Epoxy | Soft flowing | Encapsulating electrical parts |
| Silicone | Heat resistant | Encapsulating electronic parts |

Compression molding is an old and common method of molding thermosets but compared to other processes like extrusion, injection molding and blow molding, production volumes are low. Due to advanced technologies, compression molding can now be used for processing other polymer resins like thermoplastics, elastomers and natural rubbers. By this method, plastic raw materials can be converted into finished products by simply compressing them into any desired shape by using molds, heat and pressure. Unlike other processes, where molds are made up of steel, molds for compression molding are generally made of aluminum which can be easily heated. While the outside of the molds can be protected with jackets of a stronger material, the inner cores of a multicavity mold may need cavity inserts to prevent wear and thus, enable longer molding cycles with the same mold.

A basic system will consist of a press with heated platens or preferably heated molds. A two-part mold is clamped onto the two surfaces of the platens. The "female" part of the mold, when using a compound, is usually mounted on the lower platen of the press, while the "male" half is mounted on the upper platen of the press with precision guide pins to fit exactly to the bottom one. If a plastic impregnated material is used (sheet, mat), the female cavity is usually mounted on the upper platen and the male cavity at the bottom.

The polymer resin to be molded is first weighed and then preheated or the material drawn from a preheating oven and a weighed amount is then introduced into the open bottom cavity. A vertical movement of the upper platen will then close the mold allowing the heat in the mold to melt the material and flow and fill the mold completely. Strategic vent holes in the mold.

design will allow any excess gas to vent out. Here, the weight/volume of the resin compound put into the mold becomes important as too little will give a "short" product or too much material will produce "flash" and will have to be trimmed later and form wastes. After sufficient curing time and cooling, the upper platen will open and the molded product removed either manually or automatically. In a modern high-speed compression press, all operations will be are done automatically.

The preheating, mold heating temperatures and mold pressure will naturally vary, depending on the polymer resin used and mainly on their thermal and rheological properties. For a typical compression molding thermoset, preheat may be around 200 °F with mold heat and pressure around 250–350 °F and 1,000–2,000 psi. In order to ensure a full fill of mold cavity/cavities with material, some processors will use a slight excess of material which will finally end up as flash. Different methods are used to remove this flash, such as filing, sanding or tumbling. There are also systems where molded parts with flash are frozen with dry ice making it easier to deflash.

### 8.2.4.1 Transfer Molding

A transfer molding process combines the principle of compression and transfer of the polymer used for molding. In this process, the required amount of polymer charge is weighed and inserted into a transfer chamber before the molding process. This transfer chamber is heated by heating elements above the melting temperature of the polymer. The now liquid charge is gravity filled through a sprue into the mold cavity. A "piston and cylinder" arrangement is built in the transfer chamber so as to push the liquid polymer under pressure into the closed mold which is clamped. After sufficient time for a full cure, the mold opens and the molded part is removed by ejector pins or other instruments. The sprue and the gate have to be cut off.

### 8.2.4.2 Applications

This process is widely used to encapsulate items such as integrated circuits, pugs, connectors, pins, coils, studs or other industrial parts. It is suitable for molding with ceramic or metallic inserts which are placed in the mold cavity. When the heated polymer flows and fills the mold it will bond with the insert's surface. Transfer molding is also used for manufacturing large parts such as radio and television cabinets and car body panels.

### 8.2.4.3 Molding Composites by Compression Molding

The use of composites is both exciting and advantageous in actual practice, especially in the industrial sector. It works in two ways – wider range of applications and cost-effectiveness. There are many different fabrication processes used to create standard and more advanced composite parts. For example, TenCate/CCS composites who specialize in advanced composites use compression molding processes to mold parts that are very small or over 5 ft (1.5 m) in length, with thicknesses as thin as 0.050 in and well over 1 in (25 mm), using both thermosets and thermoplastic polymer resins.

In general, there are four key steps in compression molding. First, in order to mold a part, a high-strength metallic tool (mold) is fabricated, first by machining and then other finishing processes. The machining process will take into consideration and allowance for shrinkage. This mold will consist of two halves with one having a core and the other a cavity. When the mold is closed, the gap between the two halves will be the volume/thickness of the part to be made. This tooling is installed on a press, generally with vertical movement, and integrally heated and properly vented to allow gas to escape. For large parts, the cooling system should be uniform to avoid warp due to uneven cooling. Any inserts can be placed inside the mold and some processors may like to have their brand name or anything else inscribed on the inside of the mold which will appear on the outside of the molded part.

The following process sequence is more or less a standard procedure where the material is introduced into the mold with the heated material flowing to fill the entire cavity and then the cooling and holding cycle, before ejection. Here, the author would like to mention under the caption that composites are *reinforced plastics*, there are many new alternatives that can be used for reinforcing such as biomass and other wastes. Some of the items with very high potential (used in actual practice) are *bamboo fiber, rice hulls/flour, coconut fiber, graphene, wheat hulls/powder* and many others. The forthcoming book titled *Non-Traditional Fillers and Stiffening Agents for Polymers* to be published by De Gruyter will give an in-depth knowledge of this aspect of polymer processing.

While the building construction and the automobile industries use a large volume of composites, large volumes of advanced composites are also used by the aircraft industry and for space travel. According to current trends, there is increased interest in the oil and gas industries for these advanced composites.

### 8.2.5 Dip Coating

This is an interesting process in the field of plastics. Dip coating and dip molding has been around for over 100 years and over the years has become increasingly popular method of coating plastic powders and plastisols (liquid form). These processes add functionality, color, texture, electrical insulation, protection from corrosion and comfort for a variety of parts required by different industries. Despite this aspect of plastic processing, many people are not aware of the importance that dip-coated parts play in actual use.

During the early part of the twentieth century, there was a shortage of natural rubber as the demand began to outpace the available materials, PVC was created as a response to this shortage. While PVC became very popular and was in great demand, being used for plumbing pipes and fittings, the discovery that adding a plasticizer to PVC would make it flexible paved the way for the processes today known as – dip coating and dip molding. The "vinyl plastisol" (PVC with plasticizer) quickly became a popular coating for many applications.

#### 8.2.5.1 Dip Coating Versus Dip Molding

Both are essentially the same in that both techniques make use of a dip in a plastisol or powder to from a skin on the part being dipped. Dip molding creates an entirely new product with plastisol. In dip molding, a preformed mandrel or mold is dipped into a plastisol and extracted from the dip tank. It then goes through post-heat treatment and cooling processes. Finally the dip-molded product is stripped from the mold.

Dip coating improves an existing product by coating it in a plastisol to from a skin. The thickness of the coating can be determined by the number of dips and another option is dipping in additional tanks so long as the coating materials are compatible.

#### 8.2.5.2 Dip Coating Procedure

Dip coating with plastics is a process that involves several steps. The basic principle is the coating of a metallic surface/part with a plastic coating to achieve a protective layer to prevent corrosion, comfort for handling, durability and also provide aesthetic finishes since the coatings can be colored with gloss or matt finishes. These coatings are applied by:
1. Hot-dip coating in a fluidized bed of polymer powder
2. Hot-dip coating a product in a vinyl Plastisol
3. Spraying polymer powder onto a heated product

### 8.2.5.3 Hot Dip Coating in a Fluidized Bed

Metals that are to be plastic coated must be cleaned thoroughly to remove all traces of dirt, grease or foreign matter. This is usually done by using a chemical and degreasing agent. If the powders are hygroscopic, they must be heated to get rid of the moisture. The cleaned part is placed in an oven and heated to a temperature that will melt the polymer powder to be used. This temperature could be between 250 and 400 °C, depending on the polymer powder to be used.

When the metal part or parts are hot enough, they are quickly transferred to a fluidized bed that contains the polymer powder. In automatic systems, for example, coating continuous foils, these functions are done in sequence to preset parameters. If more than one coating is to be done, there will be multiple stations of fluidized beds. A compressor blows air through a fine gauze/mesh into the polymer powder, making the fine polymer particles rise and bubble within the fluidizer tank. The hot metal part is immersed in the fluidized polymer tank until a layer of polymer has melted and adhered to the surface of the metal and then removed and cooled. If the coated part has an uneven coating or insufficient thickness, it can be reheated and coated again. The resulting coating should have a fine colorful finish, free of blemishes like fine pinholes or other defects. Some of the common polymer powders used are PVC, nylon and PE.

### 8.2.5.4 Hot Dip Coating in Vinyl Plastisol

Plastisol is a vinyl compound that consists of PVC particles suspended in a liquid plasticizer. It is liquid at room temperature but forms a rubbery PVC material when heated to about 177 °C.

The process of coating with plastisols involves:
- Thorough cleaning of the part to be coated
- Heating the metal part between 175–200 °C in an oven
- Preparation of a suitable plastisol in a tank
- Placing the heated part immersed in the plastisol
- Leaving the part in the plastisol bath until the metal has cooled and does not "pick up" any more plastisol. The thickness of the partly cured PVC coating depends on the amount of plastisol converted into PVC by the heat of the metal. If the metal part cools quickly, a thin coating will be formed. If the metal cools slowly, the coating will be thicker.
- Reheating of the coated part to 175–200 °C to ensure that all the PVC cures

fully and also forms an even coat.

### 8.2.5.5 Polymer Flock Spraying

Since it is not practical to place very large products or parts into a fluidized bed of polymer powder or into a tank of liquid plastisol, they are heated and sprayed with polymer powder (flock) to achieve a plastic coating. This process involves cleaning of the metal parts using chemical cleaning and degreasing agents, then heating to a suitable temperature that will melt the flock which will be sprayed onto it.

The surface of the plastic coating may have partially melted flock stuck to it. To fully melt the flock, the product must be reheated. Repeating of the process will enable to reach the desired thickness of the plastic coating.

## 8.2.6 Vacuum Forming

Vacuum forming or thermoforming is one of the oldest methods of processing plastic materials. Vacuum formed products are all around us today and play a major role in our daily lives. This is probably the only method available in plastic processing to manufacture very cost-effective large-volume productions at high speed as especially needed by the food industry. Take, for example, the millions of thin-walled disposable food trays and boxes needed by the food chains, airline supplies and others. A simple process of heating a plastic sheet and applying a vacuum to suck in the hot sheet into the mold manually has now become a fully automated process due to modern technology.

Combining vacuum forming with new technologies such as 3D printing for toll design, thermoformable ink technology and in-mold decoration brings new possibilities as employed by some of the world's largest manufacturers. Ongoing research, for example, by Formech who collaborate with DuPont, BASF, Cambridge University and many more international technology centers to conduct research on new advanced composite materials which will bring new possibilities for vacuum forming for high-tech applications.

### 8.2.6.1 The Vacuum Forming Process

Vacuum processes offer several processing advantages over other molding processes. A big advantage is that only low pressures are used for enabling comparatively low-cost tooling (molds). Since this process needs only low-pressure operation, the molds can be made of inexpensive materials, and mold fabrication time can be reasonably short. Therefore, even prototypes and small quantity runs are economical.

Vacuum forming uses extruded plastic sheet and a secondary process may be required to trim the formed sheet. The clamp frame needs to be sufficiently powerful enough to ensure that the plastic sheet is firmly held during the forming process. Generally, it can handle up to a 6 mm thick sheet with a single heater and up to about 10 mm with twin heater machines. In automated machines, efficient safety guards must be provided to protect the operator/operators at all times.

Heaters are generally infrared elements mounted within an aluminum reflector plate. In order to obtain the best vacuum forming results using any material, it is essential that the sheet is heated uniformly over the entire surface and thickness. In order to achieve this, it is necessary to have a series of zones that are controlled by energy regulators. Ceramics do have some disadvantages in that their high thermal mass makes them slow to warm up and will also slow in their response time when adjustments are made.

More sophisticated heaters such as quartz heaters are available which have less thermal mass enabling shorter heat-up time. Pyrometers are generally used to accurately control temperatures by sensing the melting temperature of the sheet and interacting with the operating process control. With advanced machinery, computer controlled precise temperature control is possible with regular printouts for better control. Twin heaters are recommended when vacuum forming with thicker materials as they assist in providing more uniform heat penetration and faster cycle times.

Twin quartz heaters are advisable when forming with high temperature materials with critical forming temperatures. By close and accurate control of areas of heat intensity, heat losses around the edges caused by convection air currents and absorption from clamp areas can be fully compensated and a photoelectric beam is activated under the sheet of plastic during the heating cycle. Once the plastic material has reached its forming temperature or "plastic" state, it can be prestretched to ensure even wall thickness before vacuum is applied. Efficient vacuum, air pressure and optional processing aids such as a plug assist are then used to assist in molding from the heated and stretched plastic.

Plug-assisted vacuum forming is used when straight vacuum forming is unable to distribute the thermoplastic sheet evenly to all areas of the mold. To help spread the hot sheet more evenly, a device known as a plug is used to push the sheet into the mold before vacuum is applied. This method enables more of the thermoplastic material to reach the bottom of the mold and thus prevents the material from thinning out when filling the corners of a mold.

A vacuum pump is used to draw the air trapped between the sheet and the mold. Vacuum pumps can vary from diaphragm pumps to dry and oil-filled rotary vane pumps. With larger machines, a vacuum reservoir can be used in conjunction with a high volume capacity vacuum pump. This system will enable high-speed molding.

Once formed, the plastic part must be allowed to cool sufficiently before release. This parameter will be preset. If released too soon, deformation/warp will take place resulting in a reject. To speed up the cooling cycle high-speed fans can be used. Alternatively, a spray mist option using a fine spray of chilled water or other can also be used. If used in combination, the cooling period can be increased by about 30% and will also prevent possible shrinkage. Some lines will accept the trimmed waste which can go straight into the hopper/ feed station, so long as there is no contamination. Here, size reduction has to be considered.

### 8.2.7 Direct Coating

This can take the form of many processes. Here, the author will present the manufacture of artificial leather using a PVC coating by the *direct coating method*. These products are generally made according to some international standards such as BSS, DIN and ASTM .

This process involves the direct application of a prepared PVC mix onto a woven cotton cloth as the base. Base ingredients for the PVC mix are polymers, filler, color, lubricant, stabilizers or others as desired. The basic machinery needed will be a coating machine with variable speed, a "doctor knife" arrangement which determines adjustable coating thicknesses, a suitable heating oven with infrared heaters and a finishing station with an embossing roller at the other end. For preparing the PVC mix, basic machinery needed will be a triple-roll-mill, a planetary mixer, weighing machines and so on.

For ease of making the PVC mix, a processor would first calculate the color concentration needed for each manufacture and mix the filler with different colors of pigments, assisted by a small portion of plasticizer on the triple roll mill and after obtaining a fine paste keep it in different containers as color "stock." Once a formulation is worked out, the weighed ingredients, generally a 100:60–70 parts of PVC polymer plasticizer is introduced into the planetary mixer and mixed for a short period. Then, the color stock and other components are put into the mix and mixed thoroughly until a smooth liquid mix is obtained. During this process, air bubbles may form which should be eliminated.

The woven cotton cloth in rolled form is loaded onto the feed station (front), threaded between the doctor knife and the back-up roller (generally synthetic rubber) and passed through the oven and wound onto the front take up station. It is customary to now wind it back slowly to the front feed station back again. This will help the operator to remove any foreign matter, if any, and also smooth out any wrinkles on the surface of the cloth. The heaters in the oven can now be turned on low heat to avoid burning the cloth inside the oven. The first coat of PVC mix is applied with a small amount of a thickening agent as otherwise the PVC in liquid form will seep through. For aesthetic values, most processors would use colored fabric to match the color of the PVC mix.

The first base coat can be applied at higher speed than others and the heat should be such that the surface coating is only semicured to improve adhesion for the coats to follow.

The thickness of the artificial leather is predetermined and a processor will have the option of applying several coats to achieve this end. The temperature of the oven will now be raised to sufficiently cure the PVC coatings and the coatings will be done at very slow speeds. Since the coated cloth coming out of the oven is fully cured, a small heating arrangement with vertical movement is placed just above the rotating embossing/pattern roller and the postcured coated cloth will accept a finishing design. If a matt or shiny surface is required, a transparent coating without filler and color can be applied which can also act as a protective coating.

There are many systems that can be used for making direct coated artificial leather but the principle is the same. For example, a coating system can take the form of a vertical operation or a horizontal three-station operation where all coats are done in one pass. Using the same principle/methods, many other products are made by coating different substrates on different bases such as coatings on paper, cello-tape, protective coatings for steel sheets, plastic sheeting with embossed patterns as shower curtains and double-coated fabrics for protective wear, to name a few.

### 8.2.8 Indirect Coating

This method is used to make more advanced and much superior foamed artificial leather whose qualities can be a very close match for natural leather. Since natural leather is very expensive, only a few high-end car manufacturers use it and most automobiles are upholstered with the more economical foamed artificial leather available in many textures and pleasing colors.

Unlike direct coated materials, these manufactures are more complicated, need-ing sophisticated machinery and equipment. To start with, only knitted fabrics are used as the base fabric with colors matching the color of the top coat for aesthetic values. These fabrics are highly porous and thus, the need for "indirect coating" methods, meaning the coating process is a "reverse" one, where the fabric is lami-nated to the last coat surface.

The process in brief is the top coat (solid), first being coated onto a silicone-coated embossed paper of preferred design which is in roll form. This is mounted at the feed-end of the coating line where a doctor knife arrangement is also available. The basic machinery and equipment would be a feed end with arrangements for feed equipment for the knitted cloth mounted on top and the embossed paper roll at the bottom. A laminator would follow the feed-station followed by two or three heating ovens and then take-up stations for the paper and the finished coated cloth.

Here, the preparations of the PVC mixes are different in that two separate batches have to be prepared, one, Mix A (standard) and another, Mix B with blow-ing agent. Mix A will form the top coat of the artificial leather, while Mix B will be the foam layer. The functions of the various materials are: Mix A – "skin" coat or top coat, Mix B – foam layer (preferably colored to match top coat) and the knitted fabric – the base with the embossed paper giving the surface design. Some process-ors may add a top protective coating to prevent the plasticizer from evaporating due to UV light action when exposed to the sunlight.

The process: the embossed paper moving at very slow speed is given a coating with Mix A and is semicured through the ovens, rolled up and rolled back to the feed station. The foam coat – Mix B – is then applied on top of the skin coat and at the laminating station the knitted base fabric is laminated. Here, the operator will have to make sure that the surface pressure is not too much as otherwise the "wet" PVC will seep through and there must be sufficient space vertically to allow the foam coat to expand when going through the ovens. The first oven will activate the blowing agent in the mix and the second oven will fully "blow" the PVC, forming an even foamed layer. As the paper base carrying the coated cloth slowly moves forward and emerges out of the final oven, the coated cloth is peeled off the paper base and rolled up, while the paper is rolled up separately. Some machines may have autoedge trimmers but this operation can also be done later. When the coated foamed cloth is wound up, caution must be exercised to see that the possibly still warm foam is not "crushed" and the foam quality is not affected.

If, instead of PVC, PUR mixes are used for coating, the product quality will be much better in that the final products will have a finer texture and "feel." However, PURs are more expensive and also processing will be a little more difficult.

### 8.2.9 Rotational Molding or Rotomolding

Rotational molding, also known as rotomolding, is a forming process for plastics to create and produce hollow products. These can be from small to very large products such as large bins, crates, water tanks and many others. Overall, this segment of molding activity is considered the smallest in the plastic industry but one may give thought to the fact that this is the process that can make large hollow parts such as water tanks, children's playground items and so on.

This process involves preheating of a measured quantity of a polymer introduced into a very large hollow mold. The rotating action of the mold under low pressure makes the hot plastic to flow smoothly against the walls of the mold, thus forming a solid hollow part. The thickness of the part depends on the part design and the amount of polymer material used. Once the polymer has fully melted and formed the part inside, external cooling of the mold takes places with the mold still rotating. This cooling cycle will enable the formed part to shrink a little, thus enabling the operator for easy removal of the part. A molder may also use a mold release agent for easy part removal.

Although this process has a longer turnaround time than other processing methods, rotational molding is often far less expensive and is suitable for long term or short-term production runs. Rotomolded parts are extremely strong due to being seamless and with uniform wall thickness. Common polymers used for rotomolding are PEs and plastisols, while others like nylon, PP and other materials are also used for certain products.

The rotomolding process allows the use of several molds simultaneously. Specific additives can be incorporated into the polymers to enhance its final properties. Postmolding operations may include adding handles, plates and logos. Some of the rotational molding products are air ducts, water, chemical and oil tanks, security housings, small plastic boats, machine housings and many other products.

### 8.2.9.1 Solar-Powered Rotomolding Units

Thinking "green" and using latest technologies, roto-molding machines manufacturer – Light Manufacturing LLC has launched its new Solar Rotational Molding (SRM) series of solar powered units. Available as two models, the SRM-1 is a one-chamber molding unit, while SRM-2 is a two-chamber unit. This proprietary technology is designed to enable low cost manufacturing of large rotomolded products such as water tanks, boats and other items without the use of petro-based fuels.

These machines' unique designs allows them to be installed on any type of undeveloped land needing no sophisticated factory buildings, concrete pads or power from energy grids. The company's hi-tech solar power system consists of an array of "HI" heliostats which are mirrors that track the sun and reflect the sunlight onto a central point and a photovoltaic array on the roof provides energy to melt the polymer resins in the mold and rotate the mold. This energy source will also enable to operate other related equipment. This ease of installation and cheaper energy costs means that processors or customers can be very competitive in the marketplace.

The large SRM-2 comes complete with 20 heliostats and can make products as big as 650 gallon water tanks. These machines reportedly can use any roto-molding resins, including PP, PE and nylon, and manufacture the same quality products as with traditional powdered resins.

## Bibliography

[1]   Piper Plastics Corp., "Definite Guide to Dip Coating", www.piper-plastics.com 2017/09/29.
[2]   Formech, "About Vacuum Forming", https://formech.com-about-vacuumforming.
[3]   Gabriel, Jason. TenCate/CCS Composites-Article, "Compression Moulded Composites", www.azom.com/, article ID 10665.
[4]   Transfer Moulding and Compression Moulding-Lecture 4.4, https://nptel.ac.m/courses.
[5]   Science Direct, "Compression Moulding", https://www.sciencedirect.com/compression-molding.
[6]   Blow Molding, https://www.custompartnet.com/wu/blowmolding.
[7]   Wilmington Machinery, "Structural Foam Injection Molding Process", www.wilmingtonmachinery.com April 12 2010.
[8]   Milacron Marketing-LPIM (low Pressure Injection Molding), Structural Web Injection Molding, https://www.milacron.com September 20, 2018.
[9]   Milacron Blog, "What is Injection Molding?", https://milacron.com-mblog/2018/03/22.
[10]  Canadian Plastics Magazine, November 2018, www.canplastics.com.

# Chapter 9
# Processing of Liquid Polymer Foaming Systems

## 9.1 Foaming Liquid Polymers

There are many polymers in liquid form being processed into different foam products and here three selected polymers in liquid form are presented, although polyurethanes (PURs) have been left out as the author has two publications on PURs.

### 9.1.1 Polyvinyl Chloride Plastisol

Plastisols are dispersions of paste polyvinyl chloride (PVC) resins in plasticizer. Being liquids, plastisols allow low pressure or pressureless molding at room temperature. There are special grades such as copolymers and combinations, which are used for plastisols depending on the end application. Plastisols fuse or gel when heated and turn into a homogenous melt at about 150–210 °C. Preferably liquid stabilizers, paste can be mixed easier but new solid stabilizer such as CaZn are also widely used for plastisol processing. Contrary to other PVC processing methods, the required amounts of stabilizer are relatively small.

Lubricants are generally not required but would be useful for coating applications. In molding applications, for example – manufacture of fishing floats – release agents can be applied directly to the molds instead of being incorporated into the plastisol mix. If necessary, both heat and light stability agents can be enhanced by adding epoxidized soybean oil (or an epoxidized fatty acid ester like Baerostab LSU. Low toxicity, low odor and low volatile stabilizer systems have become more and more common in plastisol applications, for example, in floorings, wall coverings, toys and automobile interiors.

Some of the products made from plastisols are:
- Artificial leather (solid and expanded)
- Car underbody sealants and seam sealers
- Carpet backing and heavyweight coatings
- Conveyor belts coatings
- Dipped goods: gloves and boots
- Dolls, balls, toy animals, anatomic models for education
- Floorings (compact, expanded)
- Wall coverings, coatings
- Coated textiles, tarpaulins, tents
- Roofing membranes

https://doi.org/10.1515/9783110656152-009

### 9.1.2 Low-Density Polyethylene Foam

Polyethylene (PE) is probably the most commonly used polymer resin in the world of plastics and is made up of ethylene monomers linked together. When they are polymerized, they bond together forming a stable thermoplastic. Different polymerization methods will yield different types of PE. Therefore, PE traits are highly dependent on the polymerization process and can be divided into several categories depending on properties such as density and branching. Probably, the most common and used category of PE is *low-density PE* (LDPE).

Like basic ethylene, which is highly chemical resistant, LDPE possesses similar properties as a semirigid structure. However, PE foams can have varying textures such as soft as polyurethane foam or as hard as some polystyrene foams. There are two general types of LDPE foams distinguished from one another by the type of process used to make them.

*Extruded LDPE foams* are produced by a continuous extrusion process. The PE material is first melted until molten and a foaming agent is introduced into the mass. This mixing is under high pressure and fed into a controlled heated extrusion screw and comes out of a die opening. As the extrudate comes out of the die, generally in wide sheet form, the pressure change due to the atmosphere will cause the gas within the mixture to expand still further as the product cools and solidifies. Controlling devices can be used to ensure even thickness across the width of the foam sheet. They can be in natural white or colored and wound up in roll form.

*Cross-linked LDPE foams* – The cross-linking process can be continuous or done in batches. In chemical cross-linking, PE foam is created in batches. A solid chemical foaming agent is subjected to a temperature that causes it to blend with solid PE. This blended mix is then exposed to a still higher temperature to induce cross-linking between the foaming agent and the PE. Once cross-linking has been achieved, the temperature is raised again to induce foaming. Radiation cross-linked PE foam is manufactured via a continuous process.

Both extruded and cross-linked LDPEcross-linked LDPE foams bear high resemblance to one another with the primary difference being in the cell structure – cross-linked foams having smaller cells and softer and somewhat more uniform.

*Applications* – they have wide range of applications because of their desirable properties such as water resistance, chemical resistance, energy absorption, insulation properties, buoyancy and cushioning characteristics. Compressive strength will be greater in denser foams and naturally decreases as the density decreases. LDPE foams tend to exhibit more compressive "creep," meaning they become less thick than high-density foams.

Electric materials often use LDPE foams because they do well in applications that call for dielectric strength and constancy. Any applications that must perform in water also depend on PE foams for its high water resistance and buoyancy. The packaging industry uses this material widely to protect products because it absorbs energy and can also provide high levels of cushioning and also for its cost-effectiveness. Although these materials are chemical resistant, long exposure to sunlight may cause some basic degradation. This can be countered by a suitable additive or by coloring.

### 9.1.3 Integral Skin Foam

Integral skin foams consist of a two-component system, generally polyurethanes, that combines a lightweight flexible foam core encased in a thicker outer skin that is created in one single molding process. For example, Ecoflex integral skin foams from Foam Supplies Inc. are a good example of a versatile soft foam widely used for automobile interior parts.

Depending on the process and customer needs, Ecoflex outer skin layers can also feature decorative surface finishes, multiple color options, inserts and additives such as UV stabilizers and antibacterial agents. Inner core densities can range from "pillow soft" to very firm, depending on the end application or desired function of the finished product.

Elastoflex-W from BASF is a soft foam system derived from methylene diphenyl isocyanate (MDI), toluene di-isocyanate (TDI) and/or mixtures as custom/tailor-made formulations with densities from 30 to 80 kg/cu m. This foam system is lightweight and resilient, and because of their open-cell structures, they have good air permeability and ideal for applications for the furniture and automobile industries. Their versatility enables a processor to vary the texture and "feel" on different areas of the same product by using a combination of different densities in one molding.

Elastoflex-based foaming systems can be used to counter airborne noise and viscoelastic or high-insulation foams all comply with accepted acoustic requirements. Carpeting can be fully or partially coated as underlay among other applications which include public transport and air transport applications which demand strict compliance with fire safety regulations.

In 2017, The Dow Chemical Company produced a new polyurethane integral skin foam, also known as "self-skinning" and according to their reports helps lower global warming, while sustaining the comfort property and durability levels of their previous systems.

These new systems are designed with improved mechanical and physical properties and include:
- Lightweight and flexible
- Safety and durability
- Good flow and inform density distribution
- Superior skinning properties
- Extended shelf life of the fully formulated system
- Abrasion and chemical resistance

- Excellent adhesion to a variety of substrates
- A curing profile that allows for very competitive demold times
- Additives can be used for flame retardants, UV stability, color and antimicrobial needs

Integral skin foams are used in many applications including: automotive interiors, furniture components, household leisure goods and health-care products.

### 9.1.4 Direct Molded Foam

Australian Urethane Systems – Austhane FF Series offers a range of polyurethanes that allow flexibility in processing and create cost-saving and durable finished products for different direct molding techniques. These versatile two-component systems can be used by pour-in-place by either a machine or hand mix. These systems include various densities with special additives like fire retardants.

High resilience grades are generally used for soft molded foams in automobile interiors, seat cushion armrests, headrests and many other applications including office furniture. Specialty applications may include transport and aviation.

## Bibliography

[1]  Dow Integral Skin Polyurethanes, www.com./en-us/polyurethane/integralskinfoam.
[2]  BASF Polyurethane Flexible Foam Elastoflex. www.polyurethanes.asiapacific.basf.com. 05/05/2017.
[3]  Integral Foam Supplies Inc. Integral Skin. http://foamsupplies.com/products/integral-skin.
[4]  http://www.ausurethane.com/flexible_foam.html.

# Chapter 10
# Specialty Polymers and Polymeric Composites

## 10.1 What Are Specialty Polymers?

Specialty polymers and resins are proprietary polymers, resins, monomers and intermediates. These are based on special curing technologies or chemistries or polymers manufactured for special applications. Specialty polymers and resins exhibit properties based on their various compositions. The discovery of graphene and its tremendous possibilities has made it possible for *graphene polymers* and other specialty polymers with a combination of graphene. Special consideration should be given to properties such as tensile strength, temperature range, viscosity and water absorption when manufacturing specialty polymers. There are different basic types such as

- **Specialty elastomers** are designed for high abrasion applications. They are based on systems such as polyurethanes (PUs), chloroprene, butyl, polybutadiene, neoprene, isoprene and other synthetic or rubber compounds. Specialty elastomers and rubber compounds are characterized by their flexibility and elasticity.
- **Specialty thermoplastics** are used in bioplastics, biocomposites and some fluid-resistant applications. Because they can be reused by heating and cooling before degradation, these polymers can be reused in many processing methods, including injection molding and thermoforming.
- **Specialty thermosets** are both used in coatings and adhesives. Cured thermoset resins generally have a higher resistant to heat than thermoplastics but cannot be reused. Some products include epoxy resins, elastomer-modified epoxy resins, liquid polymers, epoxy functional monomers and modifiers and thermoset catalysts. Specialty thermosets are cross-linked polymers that are cured by heat or a combination of heat and pressure.
- **Specialty composites** are designed for applications such as noise control, vibration damping, shock isolation and cushioning. They are in products such as acoustical foams, sound barrier materials and molded isolators. Most specialty composites are strengthened by fillers like aramid fiber, carbon graphite, fiber glass, metals or minerals. These fillers, singly or in combination, will provide unique properties to meet the needs of the end applications.

There are many polymers being used for specialty polymer manufacturing and this chapter is confined to a few important applications for the benefit of the readers.

https://doi.org/10.1515/9783110656152-010

## 10.2 Applications of Specialty Polymers

The field of polymers has been expanding rapidly over the years and as the demand for polymeric applications increases, the manufacturers have been coming up with specialty polymers targeting special applications. This field ranges from basic consumer to industrial applications and even to space travel and one particular field of interest is the automobile industry.

Apart from chemistry classes, one other field where one will come across the word "specialty polymers" is in the manufacturing industry. Constant research and development has been enabling the polymer manufacturers to widen their range of these polymers and, in some cases, further improving properties as on-going projects.

There are several companies manufacturing these specialty polymers and one such manufacturer is PolyVisions Inc., which produces different kinds of specialty polymers to cover the following fields:
- transportation;
- civil engineering;
- food packaging;
- food processing;
- textiles and clothing accessories;
- healthcare and medicine.

## 10.3 Graphene – Properties, Processing and Applications in Brief

Graphene has been "discovered" as a wonder material with so many possibilities. Graphene is a two-dimensional monoatomic thick building block of a carbon allotrope, which has made a tremendous impact on industry due to its remarkable physical, chemical and electrical functionalities. Graphene and its derivatives are highly potential nanofillers that can dramatically improve the performance of polymers and polymer composites. The mechanical, thermal and electrical and other important properties such as high strength also contribute in a big way to make standard polymeric composites more efficient and usable for most high-tech applications. One of the advantages of graphene is its lightweight-to-strength ratio which is very useful for the automobile industry. Graphene polymers are also a rapidly expanding field with many possibilities.

## 10.4 Graphene–Polymer Composites

These are very versatile materials which could be specially manufactured to meet specialty end applications. New composites are being used for better food packaging, lighter cars and airplane parts among the other many applications. Over the years, researchers have been able to make tough, lightweight materials by spreading a small amount of graphene, a single layer of flat graphene sheet (carbon atoms) in polymers. In addition to being tough, these materials conduct electricity and can withstand much higher temperatures than polymers alone.

Polymers can be infused with carbon nanotubes. Traditionally, carbon and glass fibers have been used to strengthen polymers with fiber glass as a common example. Unlike with fibers, a very small amount of nanoparticles – probably less than 2% of a composite's volume – is sufficient to make a polymer stronger and heat resistant. If less filler is used, composites can retain the polymer's stretchability or degree of transparency. However, researchers have found that graphene has much better properties and is much cheaper than single-walled nanotubes which are very expensive. Graphene might also raise fewer toxicity concerns than carbon nanotubes. Carbon nanotubes could be similar to asbestos fiber and be harmful to human bodies, whereas graphene being large in two other dimensions cannot penetrate a human body and harm any of the human organs.

Graphene-polymer composites are ideal for making lighter, more fuel-efficient automobile and aircraft parts, as well as lightweight gasoline tanks and plastic containers that keep food fresh for weeks. They could also be used to make stronger wind turbine blades, medical implants and sports equipment. Because they are also good electrical conductors, they could be used for efficient and cost-effective coatings for solar cells and displays.

Clay nanoparticles and carbon nanotubes are strong contenders for use in polymer composites. For example, Toyota makes some engine parts from clay-nylon composites, which are strong and can handle much higher temperatures than nylon alone can. Carbon nanotubes infused polymers are used by producers of sports goods such as baseball bats, golf clubs and also in car parts like handles and fender parts. However, the high costs of nanotubes have restricted their use.

One method of evenly spreading the graphene in a polymer is to disperse the graphene in one solvent and to dissolve the polymer separately in another. The two solutions are then mixed together until the graphene is evenly dispersed throughout the polymer and then the solvents are evaporated. According to researchers, the use of 1% of graphene in a polymer composite showed much better results than the same amount of nanotubes used in another similar composite.

The benefits of using graphene composites in applications other than in the automobile, building construction, aircraft and others are making efficient fuel tanks and food packaging. Researchers have found that although gas and liquid molecules can penetrate through plain polymer barriers, graphene composites can provide an impermeable barrier. This means in fuel tanks, linings would keep vapor in and dissipate static electricity. Food would keep longer in packaging, since the composite material would form a barrier which oxygen cannot penetrate.

## 10.5 Graphene Composites Can Reduce Atmospheric Pollutants

Researchers have found that a combination of graphene and around 70% of titanium dioxide are more effective at removing pollutants from exhaust fumes than conventional titanium dioxide alone. Scientist from Cambridge in collaboration with counterparts from Italy and Israel have come up with a special composite to counter the tremendous amounts of exhaust fumes which is generally released into the atmosphere. This is a "photo-catalyst," which means it is activated when exposed to sunlight and is a composite made of graphene and titanium.

Exhaust fumes pollution is an increasingly serious problem, particularly in cities and developing countries. Probably, the most troublesome sources of gases that are highly hazardous to human health come from vehicle exhausts and thus highly congested cities experience high levels of pollution. One theory is that if these composites are applied to the cement surfaces or cement being used for buildings and other infrastructure, to clean the air as the sunlight strikes the surface. According to reports, this new composite is 70% more effective than titanium alone, and tests carried out against rhodamine B, a compound of similar structure to volatile organic compounds (VOCs) also found in air pollution, were 40% more efficient than titanium alone.

Photocatalysts in a cement matrix applied in building construction could have a large effect in decreasing air pollution by self-cleaning of the surfaces termed as the –"smog-eating" effect. Graphene could help improve the photocatalytic behavior of catalysts like titanium and also enhance the mechanical properties of cement.

## 10.6 Fluoropolymer Specialty Insulation for Extreme Application Needs

Fluoropolymer composite materials play an important role as wire and cable insulation. Especially in high-voltage systems in high temperature or harsh environmental applications, their roles are even more crucial. For example, in aerospace applications using a film with inadequate dielectric strength or temperature resistance can cause the insulation to fail, bringing down an entire electrical system. Extreme, critical and harsh environmental conditions call for high-performance insulating materials.

Take for example the FLUOROWRAP products which are specially engineered composite materials which are available in a wide range of extruded polytetrafluoroethylene (PTFE) tapes as well as heat-sealable polyimide fluoropolymer composite tapes. All products feature outstanding dielectric strength, estimated up to 7,000 V/mil for high-voltage power cables. They also feature excellent resistance to harsh chemicals, heat and fire.

Their products are especially suitable for demanding applications like
- **Unsintered extruded PTFE:** Meets a number of military standards for PTFE wire for aerospace power cables, coaxial cables, thermocouple wires, battery membranes and others.
- **Sintered extruded PTFE:** Has excellent permeability resistance for high-temperature wire as well as magnet coil windings.
- **Low-density extruded PTFE:** For high-frequency coaxial cables.
- **Cast PTFE:** With heat-sealable bonding and laser-markable options for use in extreme environments requiring laser marking or heat sealing.
- **Polyimide fluoropolymer composites:** With high-voltage dielectric resistance, these are ideal for most demanding applications.

## 10.7 Polyurethane Specialty Polymers

Specialty polymers such as polyester diols, thermosetting and thermoplastic polyurethanes (PTU) based on esterification technology are used in binders, adhesives and other applications requiring solution PUs and TPUs. Super absorbent polymers can absorb and retain extremely large amounts of a liquid relative to their own mass.

TPUs are high performance polymers. There are many companies who deal with specialty polymers among whom the Songwon Industrial Group specializes in PUs. Their range of TPUs under the brand name Songstomer are used by manufacturers of high-quality products such as wire and cable, film and sheeting, mobile phone are ideal for the most demanding extrusion, injection molding and calendaring requirements due to their special properties.

The following are some of these products suitable for different applications:
- **Polyurethane laminating adhesives:** Increases productivity of packaging. Excellent physical properties and other technical attributes.
- **Polyurethane ink binders:** Printing ink manufacturers can print high quality prints on plastic films by benefitting from these excellent physical properties and adhesive properties and also ideal for gravure and flexography printing inks used in flexible packaging laminates.
- **Polyurethane coatings:** Designed for a wide variety of applications like artificial leather, gloves, shoes and others. These products are distinguished by high wetting properties and low tackiness, making them particularly suitable for dip-coating processes. These grades are one-component materials that form films with excellent stability at low temperatures, whether applied by dry or wet processing methods.
- **Polyurethane adhesives for artificial leather:** The range of PU and hot melt adhesives promotes high tensile strength and resilience, and provides excellent elongation properties and outstanding resistance to chemicals, oil and abrasion. At the same time, they help to impart a soft and comfortable feel to the finished material.

## 10.8 Electrical Conductivity of Conjugated Polymers

Polymers traditionally have been good insulation materials, for example, insulation of electrical cables and electrical devices. However, there are also some polymers that are electrically conductive. The property of electrical conductance is based on the presence of conjugated double bonds along the polymer backbone. Conjugation means that the polymer backbone consists of alternating single and double bonds, although the conductivity of these polymers will be low. When an electron is removed from the valence band by oxidation or added to the conducting band by reduction (processes of doping), the polymer becomes highly conductive.

Polymers have always been considered as insulating material, particularly for electricity. Electrically conducting polymers or synthetic metals combine the electrical properties of metals with the advantage of polymers to offer great properties like lightweight, greater workability, resistance to corrosion and chemical attack, lower costs having applications from consumer goods to space travel, aeronautics, electronics and nonlinear optics. The discovery and availability of these special materials have opened up many new possibilities. Among some of these novel conducting polymers are polypyrrole, poly phenylacetylene, polythiophene, polyfuran and polyaniline and their derivatives.

Planarity and large anisotropy ratio (interchain conductivity) will have a wide range of conductivity depending on the degree of doping, the alignment of chains, the conjugation length and the purity of the sample.

### 10.8.1 Flexible Conductive Film

Researchers at Purdue University – Ind. have designed a bendable polymer to conduct electricity for transparent and flexible electronics. This specialty polymer is made from long chains that contain radical groups which are molecules that have at least one unpaired electron. This polymer film has the look and feel of glass and can be inexpensively and sustainably produced on a large scale because it comes from earth-abundant materials. This material is reportedly cost-effective compared to currently used polymers in electronics that rely on expensive chemistry and chemical doping to achieve high conductivity.

When produced commercially, these conductive films will be much cheaper than the current – indium tin oxide – which is the current standard material used. These films come under the category bioelectronics and it is expected that with further research these films will lead to valuable biomedical products like noninvasive sensors for medical applications.

### 10.8.2 Some Latest Automotive Specialty Polymers

Due to increasing environmental concerns and rising production costs, automakers have been always on the lookout for auto components that are tougher, lighter, more corrosion resistant, than traditionally used metals.

Over the years, specialty plastics have been replacing metal parts used in automobile manufacturing but as new challenges rise up, there is a constant demand for more and improved components. Lanxess AG has come out with a new product range called Durethan Performance for periodically stressed components. According to them these new grades are several times more resistant to fatigue under pulsating loads than the standard glass-fiber filled content. Their initial product range are the thermally stabilized Durethan BKV30PH.2.0, BKV35PH.2.0 and BKV40PH2.0 compounds with glass-fiber contents of 30%, 35% and 40%, respectively, as well as the impact resistant-grade Durethan BKV130P.

Marketing has always been a challenging, although an exciting aspect of manufacturing. For automakers, the buyer's expectations of a "nice smell" to promote the concept of a "new car" were a bonus factor. However, over the years, this has changed in that the buyers tend to associate this nice smell with harmful emissions, including VOCs. Therefore, the objective now is to get rid of this "smell" out of the new cars. To meet this need, PolyOne Corporation has introduced a low odor talc-filled PP called Maxxam LO to meet the vehicle interior required air quality. There are different grades, which produces a neutral-smelling interior according to them and they can be also customized.

Struktol Company of America is offering a line of products targeting odor and VOC control with their new Struktol RP 53 additive. This specialty product uses a blend of chemistries to neutralize high-odor compounds containing problematic components such as mercaptans, amines and phosphites. This product can be used in a variety of polymer resins and will be an effective neutralizing agent for automotive car interiors.

## 10.9 Polymer Modified Bitumen

Bitumen is a by-product of the fractional distillation of crude oil but is also found in natural deposits. From ancient times, bitumen has been used for waterproofing and as an adhesive and it is a low-cost thermoplastic material but brittle in cold environments and softens readily in warm environments. One of the many methods to toughen bitumen is to blend it with either virgin or scrap polymers to produce a better product – *polymer modified bitumen.* In this modification, the author suggests the addition of 10–12% rice hulls or wheat hulls ash that contains high levels of silica, which will enhance the PMB properties further with added moisture barriers and strength due to the presence of silica.

## 10.10 Reinforced Polymers with Coffee Chaff

Innovative Canadian research by the University of Guelph has made it possible to come up with over 60 million pounds of coffee chaff – the unused outer skin of the coffee beans during the roasting process – which would normally end up in land-fills. In partnership with the inventors, Ford Motor Company, USA, is now convert-ing these wastes into new modified durable polymers which goes into reinforcing auto parts, including some which needs to meet the highest demands.

The process includes treating the chaff with high heat and mixing plastics and suitable additives and turning the material into pellets. These can be molded into various auto parts like headlamp housings, battery covers and other interior and under the hood parts. Two important aspects of these materials is that they are about 20% lighter and requires 25% less energy to mold than materials that have been used earlier. This new resins should have great potential with a wider range of applications.

## 10.11 Specialty Polymer Composites

Composites are materials made with two or more constituent materials with signifi-cantly different physical and chemical properties that if combined, will produce a material with characteristics, different from the individual components used, while remaining separate within the structure. A composite material can also be described as a macroscopic combination of two or more distinct materials having recognizable interfaces between them, the properties of which can be optimized by the addition of additives to achieve a balance of properties to meet the requirements of a given range of applications.

Several commercially produced composites use a polymer matrix which can be com-prise of a single polymer or a combination, so long as they are compatible, with many natural polymers and modified polymers being available. This selection is highly depen-dent upon the end requirements as well as the compatibility between the constituents to be used. Some of the most common polymers are polyesters, vinyl ester, epoxy, phenols, polyimide, polypropylene (PP), polyvinyl chloride and polyethylenes (both low density and high density). Reinforcing materials are very often fibers but also ground minerals. In polymeric composites, the polymer matrix can also be considered a "carrier."

The ratio of the polymer resin matrix to reinforcing constituents used is very im-portant just as the compatibility between all materials. As a general rule of thumb, a 60% polymer matrix to 40% reinforcing material maybe a good starting point. Some of the polymeric composites, for example, those needed by the auto industry, build-ing construction and space travel will call for highly technical properties which the composites will have to meet. Here, two basic properties like strength and durability may take priority. It is possible to reduce the content of the polymer matrix but there is a limit to this aspect as the overall strength of the final composite material will depend on it. Too little polymer resin may cause problems such as blending and cohesiveness as composites are made with a wide range natural fibres and biomass. These emerging family of composites can be categorized as "*thermoplastic biocomposites.*"

## 10.12 Specialty Engineered Composites

Engineered composite manufacturing needs special technologies and are generally made to specific shapes. The matrix material can be introduced before or after the reinforcing materials are placed into a mold cavity surface as in the production of fiber-glass products. The matrix material undergoes melting and the part shape is set. Depending on the nature of the matrix material, this melting can occur in various ways such as chemical polymerization or solidification from the melted state.

## 10.13 Specialty Polymer Resins with Rice Hulls

With periodic escalating polymer prices, resin researchers and developers have been coming up with a composite resin with a polymer matrix combined with rice hulls powder/flour. The production of these composite resins and final extrusion process can be done in two stages. To achieve high quality composite resins, it is recommended that advanced processing systems be used. The first stage can be sieving to eliminate foreign matter, ground to a fine powder, dried to get rid of any moisture and then put through an extrusion process resulting in pellets. The combination can be a ratio of – polymer matrix to rice hulls powder – 40:60. Using masterbatches, these pellets can be colored as desired. Alternatively, the entire production process can be done in one operation where the pellets are fed into a hopper of an extrusion system to make profiled extrudates.

Other processing methods for these composites resins are injection molding and compression. Rotomolding is another possibility. These composite resins can be made with various polymers or combinations of compatible polymers forming a polymer matrix. Also, the reinforcing agents can be biomass such as flax, palm-fiber wastes, wood flour, rice straw and others or combinations of them, as long as they are compatible. Laboratory tests have shown that composite resins with rice hull powder – the author calls them *PCRH,* gives better overall properties such as structural strength, processing ease, easy fabrication and finishes. Extruded boards, for example, are an ideal substitute for natural wood and can be fabricated with traditional tools used for wood. Moreover, these boards can be produced with ideal wood grain veneers. Depending on the end application, a small quantity of a blow-ing agent may be incorporated since the extruded boards tend to be a bit heavy.

If colorants are not used, these composite resins will be from slightly transparent to yellow. Properties will depend on the constituents and additives used, whereas the pellet sizes can be as desired. Advantages of using these resins in injection molding and compression molding are lesser energy required for molding and shorter cooling and dwell times.

### 10.13.1 Polymeric Composites with Rice Hulls for Injection Molding

At the beginning, injection molders were cautious in their approach to these new composite resins but as they got more acquainted with them and considered the advantages of using these environmentally friendly materials, they began to appre-ciate the advantages, especially cost savings. Due to constant development of these versatile materials, composite resins with different polymer matrices have opened a whole new frontier with exciting possibilities.

Wood fiber blends had already made a name for themselves and have been in use for some time and now composite resins with rice hulls have effectively moved into injection molding. Some injection molders may still be reluctant to use these composite resins in comparison to the traditional resins they are familiar with. However, with growing environmental concerns, rising resin costs and biodegradable problems and other factors, injection molders and other polymer processors are showing greater in-terest in these new materials.

Constant research and development in the production of these composite resins have significantly improved the quality and processing ability of these resins. One big advantage of using the new generation of composite resins is that they can be easily processed on traditional processing equipment with minimal adjustment of processing parameters, with no other physical hardware needed.

These PCRHs can be made with a variety of polymers like polyethylene, polysty-rene and PP or with combinations of them, as long as they are compatible with each other. These versatile reins can be classified as an emerging family of material that can be called *thermoplastic biocomposites*. It is a huge advantage for polymer processors that these composite resins are now available in different grades and properties. In general, the processing parameters for all grades are more or less the same and very similar to wood polymer composites.

There are many compelling reasons to use alternative materials such as these composite resins as they contain around 50% organic fiber and offer cost-effective resins with better properties in most cases. For polymer processors this is a viable option, since these resins are environmentally friendly than petrochemically derived polymers. In addition to the "green" factor, these composite resins reduce a molders exposure to rising resin prices, reduce the energy costs associated with production, while producing components with greater structural strength in aesthetically pleasing finishes with high marketplace potential.

Polymeric rice hull composites tend to be lower in costs and weight than unfilled or glass-filled resins and even competitive with resins filled with calcium carbonate or talc. An added advantage is lower density and lower costs which are beneficial in applications where a premium is put on light weight applications such as transportation, automobile, sports, aviation, space travel and also consumer products.

Rice hulls are abundantly available in most countries at virtually no cost and basically consist of about 20% opaline silica and a polymer called *lignin*. From the rice milling process results in two basic grades of hulls – coarse and fine. The particle size of the hulls are naturally of importance as well as the moisture content and further grinding for reduction of particle size and drying to reduce the moisture content as suitable to manufacture composites. According to researchers, they do not flame or smolder and does not transfer heat or emit odors.

To achieve good quality finished products, it is important to use high quality composite resin pellets. There are three basic areas vital for selecting these pellets as follows:

**Moisture content:** Surface moisture should be less than 1.5%, while the internal pellet moisture content should be ideally less than 1.0%. Increased moisture contents and failure to control the moisture during processing will result in "splay" (whitish patches) and excessive brittleness.

**Pellet characteristics:** Pellets should be clean and relatively consistent in size and shape. There should be no "chads" or "streamers." Powdery residues on surfaces will indicate the use of nonstandard pellet manufacturing equipment or poorly maintained machinery and equipment by the manufacturer.

**Correct grades:** One of the benefits of composite resins is that they can be easily blended with additional virgin resins, if necessary. Blending molders can achieve different performance characteristic in addition to basics such as flame retardants, color and extra glossy or matt grades.. Thus molders will have the choice of selecting the correct grades to suit particular end applications.

## 10.13.2 Recommended Processing Guidelines

The four main processing methods for polymers are *injection molding, extrusion, compression molding and blow molding*. Rotomolding is also a process where these composite resins can be used. Here, the author will present recommendations for injection molding.

When molded properly with correct temperature, speed and a nonresistance flow path, products will exhibit minimal stress, smooth surface, uniform color distribution and no evidence of gassing. The two most important principles to remember when molding polymeric composite resins with rice hulls are to avoid excessive heat and shear. While some may think that the rice hulls in the composite will act as an inhibitor, tests have shown that actually the reverse is true. For example, PP composite resins with rice hulls flow very quickly at relatively low temperatures and pressures, as a result of which processors can achieve significant energy savings. Added features are shorter cycle times and higher productivity due to reduced fill times and cooling times.

Recommended temperature guidelines for standard processing equipment are

Rear zone: 340 to 370 F
Middle zone: 360 to 390 F
Front zone: 380 to 410 F
Nozzle tip: 390 to 410 F

Taking these as basics, molders may have to adjust these parameters when using different grades to achieve the best results. Molding pressures will depend on the part design as well as the runner system and the gates. As a principle, molding with these composite resins will require less pressure than molding with traditional resins. However, molders should be careful with filling speeds. While the hot material will tend to flow quickly, it is important to avoid excessive shot fill times as these materials are shear sensitive. If streaking occurs, it can be remedied by simply slowing down the injection time. Given that lower temperatures are used for molding, hold times (dwell times) are often much lower for standard materials.

The nozzle tip used in processing these composite resins should have an orifice as close as possible to the diameter of the sprue to minimize shearing. Smaller orifices may cause increased shear as well as discoloration caused by overheating of the material as it enters the mold. Injection molded parts with these composite resins are rather "natural" in color with a light brown tone and uniform grain. However, parts can be easily colored with high gloss or matt finishes by additives.

With an abundance of rice hulls, as a renewable biomass available and with researchers and resin developers coming up with grades with newer and improved properties, polymer processors could expect a great future, which will be both exciting and profitable. The availability of laboratory facilities and training for processors will only enhance and enable them to produce high quality parts to meet any type of end application.

## 10.14 Expanded Polystyrene with Graphite

"Terrafoam" is an environmentally superior polystyrene insulation product made from a new polymer by BASF called NEOPOR. It provides higher thermal insulation performance without any hydrofluorocarbon (HFC) blowing agents and contains graphite that does not off gas, thereby providing permanent thermal performance.

NEOPOR is a special polymer used to produce thermal insulation for the building industry. This unique material carries a distinctive platinum color due to the graphite contained within the polymer matrix of the expandable polystyrene beads. The graphite particles both reflect and absorb radiant energy, thereby increasing the insulation capacity or R-value by as much as 20%. The reduction of the radiant energy transmission occurs within the insulation itself. NEOPOR is used worldwide in insulation applications where cost-effectiveness and sustainability are priorities. Beaver Plastics, Canada, produces NEOPOR insulation products under license from BASF. Their product range includes insulation for walls, below grade, under slabs, roofs, high stress geotechnical applications and exterior insulation and finish systems.

## 10.15 Polymeric Composites with Bamboo Fibers

Bamboo fibers are one of the most important natural fibers among some of the many being used. The versatility, amazing properties, cost-effectiveness and availability in abundance makes these fibers a forerunner for polymeric composites. Natural fibers belong to the category of *renewable biomaterials* for which increasing interest is being shown by scientists, researchers and developers. Bamboo cultivation requires much less energy and resources compared to production of glass fibers and carbon fibers and bamboo fibers will be a cost-effective and ideal reinforcing agent for polymeric composites.

## 10.16 New Polylactic (PLA) grade for 3D Printing

A thermoplastic polymer derived from renewable sources like corn starch or sugar cane – PLA – has a relatively low glass transition temperature between 43 and 60°C, which makes it unsuitable for high-temperature applications. At these temperatures, even parts in a "hot" car during summer, for example, can soften and deform.

However, Minnetonka, USA, a biopolymer supplier has introduced a new PLA 3D printing grade designed specifically for 3D filament manufacturers seeking to tap into the lucrative industrial prototyping market. This special grade is called Ingeo 3D870 having better properties like impact strength and postprint annealing heat resistance. According to the manufacturer, this grade is an extension of their previous grade 3D850 and is ideal for industrial 3D printing, allowing a wider range of printing and does not need to be printed with a heated build chamber to eliminate possible warping as is necessary with industrial prototype parts printed with acrylonitrile butadiene styrene filament.

# Bibliography

[1] Tang, Long-Cheng, Zhao, Li, Guan, Li-Zhi. Article on "Graphene/Polymer Composite Materials, HangzhouNormal University, 311121 China.

[2] Bhakhshi, A.K., Bhalla, Geetika. Journal of Scientific & Industrial Research, Vol.63, Sept., 2004, pp. 715–728.

[3] Songwon Industrial Group, Article: "Tin Intermediates, PVC Additives and Polymers", www.songwon.com//products/tpp.

[4] Specialty Polymers and Resins Guide Engineering 360, "Specialty Polymers and Resins", www.globalspec.com/learnmore/materials.

[5] Stuart Nathan. Article "Graphene composite degrades atmospheric pollutant", www.theengineer.co.uk/graphene-composite-degrades 4th Dec. 2019.

[6] Patel, Prachi. Article "Graphene-Polymer Composites", MIT Technology Review, May 27 2008, www.technologyreview.com/s/41018/graphene-polymercomposite.

[7] Publication: Saint-Gobain Specialty Films, "Fluoropolymer Insulation for Your Extreme Applications", www.saint-goobain.com, Sept. 12th 2018.

[8] Publication: "Industry Applications of Specialty Polymers", Poly Visions Inc., www.polyvisions.com/specialty-polymers.

[9] Canadian Plastics Magazine – February 2019, www.canadianplastics.com.

[10] Canadian Plastics Magazine – April 2017, www.canplastics.com.

# Chapter 11
# Three Manufacturing Processes for Important Products

From Chapters 1 to 10, readers would have gained a good knowledge of processing polymers. For their benefit, the author now presents three manufactures in detail as follows:
(1) Manufacture of molded products for expandable polystyrene (EPS)
(2) Manufacture of polymeric composite resins
(3) Manufacture of polyurethane (PUR) foam mattresses

## 11.1 Molded Expanded Polystyrene Products

Some of the main applications of EPS molded products are packaging, insulation, fish boxes, pipe insulation, building construction and floating devices (fishing floats), while large extruded sheets with a "skin" is specially made for the building construction industry. The main advantages of EPS material is lightweight, high insulation factor, low thermal conductivity, very low water absorption and excellent cushioning effect which is ideal for packaging. The molding process for small volumes can be a simple process with self-fabricated machinery, except for the steam source. For medium size production operations manually operated or semi-auto machinery can be used with production of large volumes and large size foam blocks. One should invest in fully automatic machinery.

The raw material resin is in the form of tiny colorless beads ranging from 0.5 to 1.3 mm in diameter with some manufacturers offering larger size beads. These are generally available in 25 kg paper bags, larger volume steel drums or in big totes and must be stored in room temperature around 20 °C and their expected shelf life is about 6 months. Since each bead has a tiny amount of pentane gas (blowing agent) incorporated into the beads at time of polymerization, the packs must at all times be kept closed to prevent loss of gas to the atmosphere. Self-colored beads are also available, if desired but more expensive. It is not possible to color the beads on a production floor as one can achieve only a "mottled" effect at best.

There are different grades of beads offered by manufacturers such as general purpose, fire retardant, marine grades and packaging grades. There is also a special grade called water-blown EPS (WEPS) which will need different preexpanding equipment. Some of the producers of EPS raw materials are BASF, Bayer AG, Dow Dupont, Sabic, Synthos Group, NOVA Chemicals and others from countries like China, Japan and South Korea. Some of the well-known brands for EPS products are Styrofoam, Rigifoam, Thermocole and Geofoam.

https://doi.org/10.1515/9783110656152-011

The general processing operation will be as follows:
- preexpansion and aging;
- curing;
- molding process;
- shape molding;
- fabrication.


## 11.1.1 Preexpansion and Aging

The first step is to expand the beads. These beads can be expanded up to 50 times its original size but this is done in two steps: first being preexpansion (part-expansion) and the final expansion occurs when being molded into shapes.

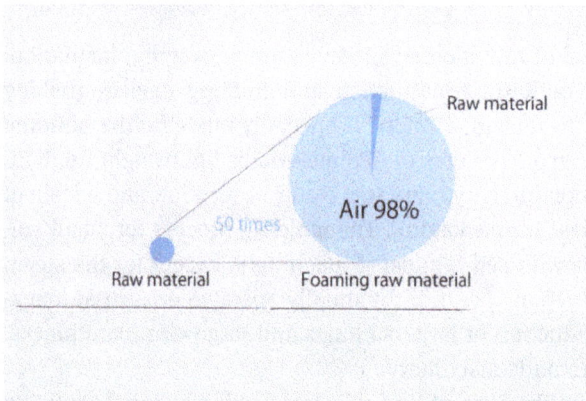

If sophisticated technology equipment is used, a computer controlled weighing system will introduce a measured amount of beads into a preexpander. Steam is introduced into the vessel and an agitator will mix the beading beads as the heat from the steam will release the pentane gas from the beads expanding them. The degree of expansion will depend on the final density to be achieved. A level indicator will tell the computer when the predetermined expansion volume level has been reached. A short drying period to get rid of the moisture present will take place using a fluidized bed before these dried beads are "blown" into large bags or silos for the aging process.


## 11.1.2 Curing Phase

These beads have been under a dynamic physical transformation that has left them with an internal vacuum in the millions of cells created. This vacuum must be equalized to the atmosphere to prevent bead collapse or implosion. Thus, the curing phase allows the beads to fill back up with air and equalize. This curing process can be from 24 to 48 h, depending on the desired final density of the material. Now, the beads are ready for molding. A note of interest here is that WEPS beads when expanded does not need a curing phase and can be molded directly.

### 11.1.3 Molding Process

The molding process involves filling molds made of aluminum with this cured loose beads and forming them into a solid mass. The moldings can take the shape of large solid blocks, for example, 2 m × 1 m × 1 m or larger. High-tech molding equipment will have state-of-the-art vacuum-assisted molds for automatic molding processes. Here, the hot molded EPS block is cooled automatically by air instead of water and ejected automatically from the mold and then another molding cycle commences. In simple terms, once the rectangular cavity is filled with the predetermined amount of beads which depends on the final density desired, the computerized system will use a vacuum system to remove the air from the material. Steam is then introduced into the mold through strategically placed steam inlets, flowing through the entire mass of beads. The heat from the steam softens the surface of the beads and as they expand further under pressure and heat, the beads will fill the voids between the beads and fuse together forming a solid mass. The computer will then release the pressure inside and the solid mass is cooled by vacuum and the solid block is pushed out from the mold. Sufficient cooling is needed before release from mold to prevent warping or unwanted expansion of sides. The next process for these blocks will be discussed under fabrication.

### 11.1.4 Shape Molding

This molding process more or less is the same as for block molding, the difference being that contoured aluminum molds are used instead of large block molds. Shape molded products are used for applications like packaging, hot/cold containers, boxes, insulated concrete forms and pipe insulation. For this operation, generally machines with platens moving vertically or horizontally are used where the molds are mounted on the platens. The molds are made of aluminum and consists of two halves, on the principle of "male/ female" contoured cavities. These molds can also have be multicavity depending on the products to be made.

   Manufacturers of molding machines offer manually operated, semi-auto or fully auto machines with water or vacuum cooling systems. One platen is stationary, while the other moves opening the mold with auto ejection of the molded part or parts.

### 11.1.5 Fabrication

This operation refers to the cutting of the large foamed blocks using horizontal and vertical cutting machines. It is customary for all sides to be trimmed first before cutting the block into desired thickness of sheets, slabs or other shapes. For this operation either hotwire systems (electrically heated wires) or band saws can be used with the former giving better and smoother cuts with less waste. For example, while it is possible to cut 10–1 cm sheets from a 10 cm block with hotwires and saw (thin), and it is not possible to do this with band saw blades (thicker). Depending on the end application, if the molded foam block is reasonably smooth on all sides, the initial trimming may not be needed. The general principle of cutting involves the movement of the foam block on the bed of the cutting machine or the movement of the cutting wire system while the foam block is stationary. Hotwire systems can have multiple wires with automatic settings enabling a cut of different thicknesses with one pass.

Since the EPS industry is well advanced, there are very sophisticated cutting and fabrication machinery which can cut various shapes and contours used as facades, decorative applications, ceiling tiles (can be molded also). Also, for some industrial or specialized applications there are many protective and decorative coatings in the market.

### 11.1.6 Some EPS Molded Products

The following products are shown as an illustration of some molded products:

Figures 11.1–11.3 show fish boxes, packaging and floats. Photos: author's factory in Sri Lanka.

### 11.1.7 Expanded Polystyrene Properties and Key Benefits

EPS material provides many excellent properties and key benefits over other similar materials and forms probably the best material for applications like insulation, packaging, hot/cold containers and decorative fabrications.

#### 11.1.7.1 Key Benefits
- Energy efficiency and cost savings
- Thermal resistance and insulation
- Chemical inertness

- Bacterial resistance
- Cost savings
- Excellent impact strength
- Excellent dimensional stability

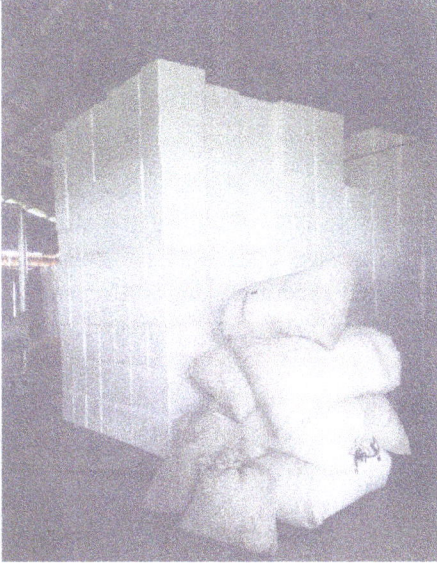

**Figure 11.1:** Fish boxes.

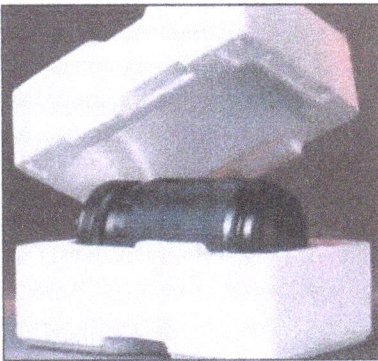

**Figure 11.2:** Packaging.

**Figure 11.3:** Floats.

### 11.1.7.2 Properties

**Thermal properties:** EPS has very low thermal conductivity due to its closed cell structure consisting of 98% air. This air trapped within the cells is a very poor heat conductor and hence provides the expanded foam with its excellent insulation properties. The thermal conductivity of EPS of density 20 kg/m$^3$ is 0.035–0.037 W/(m.K) at 10 °C.

**Mechanical strength:** Flexible production methods make EPS a versatile material in strength, which can be adjusted to suit the end application. EPS with high compressive strength is used for heavy load-bearing applications, whereas for void forming in building construction work, a low compressive strength will be needed.

As a general rule, strength characteristics improve with increase of density and for higher densities the preexpansion can start with the smaller bead range. It is also possible to mix smaller and larger beads before preexpansion to achieve a particular density. However, the cushioning characteristics for EPS foam packaging will depend on the geometry of the molded part and to a lesser extent by bead size, processing conditions as well as density.

**Dimensional stability:** EPS offers exceptional dimensional stability, remaining unaffected across a wide range of ambient factors. The maximum dimensional change of EPS is around 2% which is in accordance with ASTM D2126.

**Electrical properties:** The electric strength of EPS is approximately 2 kV/mm. Its dielectric constant measured in the frequency range of 100–400 MHZ and at densities from 20–40 kg/m$^3$ lies between 1.02 and 1.04. Molded EPS can be treated with antistatic agents to comply with electronic industry and military packaging requirements.

**Water absorption:** EPS material is not hygroscopic. Even when immersed in water, it absorbs only a very minute amount of water, which is negligible. As the cells walls are nonabsorbent water can penetrate only through any channels available through any unfused beads, which is very unlikely. For special applications, the surface of EPS can be painted or a protective layer can be applied. For example, when fishing floats for nets are made, they can be applied with a suitable protective coating in different colors in keeping with international codes.

**Chemical resistance:** Water and aqueous solutions of salts and alkalis do not affect EPS. However, EPS is readily attacked by organic solvents.

**Weathering and aging resistance:** EPS is resistant to aging. However, exposure to direct sunlight (UV) leads to a yellowing of the surface accompanied by a slight embrittlement of the upper surface. Because of the low depth of penetration this yellowing has no significance for the mechanical strength of insulation. Also protective coatings can be used.

**Fire resistance:** EPS is flammable. Incorporation of flame retardants significantly reduces and is a must to meet building construction codes. These additives will prevent the spread of fire and localize the fire hazard.

### 11.1.8 EPS in Building Construction

EPS is widely used in the building construction industry, mostly for insulation and also as decorative ceilings, facades, lost foam, paneling applications and so on. This material's special properties like lightweight, chemical inertness, pest resistance are additional reasons why this material is important. Its closed cell structure preventing absorption of water/moisture is an advantage in applications like roof insulations, floating devices, marinas and is also used in large industrial applications like road building, railway constructions and others.

### 11.1.9 Food Packaging

EPS can be extruded in very thin sheets in continuous form using conventional equipment. This is the process to make thin-walled food trays and fruit trays using a vacuum process to make any shape desired. As we know, these products are needed in very large quantities by airlines, grocery packing, food take-out places and many others.

EPS does not have any nutritional value and therefore does not support any fungal, bacteriological or any other microorganism growth. Therefore this material is ideally used for packing seafood, meats, fruits and vegetables among others. EPS helps to keep food fresh and prevent condensation throughout the distribution chain of activity. Another wisely used application is disposable cups and plates.

### 11.1.10 Industrial Packaging

EPS is the most popular material used for industrial packaging. Due to its shock absorption properties, EPS is the ideal material for use to ensure protection and safety for products from risks of transportation and handling. This rigid foam material can be molded to any desired shape and is widely used in packaging for electronic devices, TVs, sensitive products, toys, domestic goods and the list is endless. EPS is also used for domestic and industrial coolers and hot/cold containers. Different grades of beads are available and a processor will choose the best grade for different applications and often guidance from the bead manufacturer is helpful.

### 11.1.11 Safety and Recyclability

EPS material is composed of organic elements such as carbon, hydrogen and oxygen and does not contain chlorofluorocarbons or hydro chlorofluorocarbons. EPS is generally recyclable at any stage of a cycle. EPS is 100% recyclable and is designated bits plastic resin code 6.

However, the collection and storage of EPS wastes can pose a major challenge as they are very light and have large volumes. Some recyclers have found solutions on converting these wastes into new products.

### 11.1.12 Extruded Polystyrene Insulation

There are four major rigid plastic foam insulations commonly used for residential, commercial and industrial insulation. They are extruded polystyrene (XEPS), EPS, PUR and polyisocyanurate. Each one has its own characteristics with specific advantages and disadvantages for particular building applications. However, the most used and popular on is XEPS in large extruded sheet form. An added advantage is the formation of a surface "skin" during extrusion which will add to the insulation properties.

XEPS foam sheets are available in different colors such as blue, yellow and pink and each having different properties so that the builders can identify and use the correct ones for particular applications. The thicknesses and lengths will vary according to manufacturer's specifications to meet building construction codes. Some of the different properties of these sheets are densities, fire retardant factors, impact strength, antibacterial and water absorption among any others a builder may need.

### 11.1.13 The Extrusion Process for XEPS

The process begins with solid polystyrene (PS) crystals, along with special additives and a blowing agent being fed into an extruder. Within the extruder, this mixture is combined and melted under controlled conditions of high temperature and pressure, into a viscous plastic mass. This hot mass is then propelled forward by the screw and extruded through a predetermined die, generally a T-die. As it emerges from the die, the foamed material will expand and will pass through a calibration unit which will control the dimensions of the extrudate. Cooling will take place and the material in a continuous form will be slowly carried forward toward a vertical saw which will cut the foam material into desired lengths.

This continuous extrusion process will result in a unique foam product in the form of sheets with a uniform closed-cell structure, and a smooth continuous skin with insulation qualities unequalled by other insulation materials. EPS sheets cut from large EPS molded foam blocks also can be used for insulation, which probably will be cheaper but XEPS sheets are a better option.

### 11.1.14 Insulation Properties of XEPS Boards

The insulation properties of XEPS boards or sheets will be better than boards or sheets cut from molded EPS blocks. The ability of an insulation material to resist heat flow is usually the principle consideration in comparing insulation materials. A material's resistance to heat flow is expressed as the *R*-value. The higher the value, the greater the insulating factor.

## 11.2 Manufacturing Plant for EPS Products for an Entrepreneur

For the benefit of an entrepreneur, the author presents this cost-effective plant setup and manufacturing process from a hands-on experience. Based on the assumption that an entrepreneur will have limited resources, this recommendation is ideal for the processing of around 2 metric tons per month of EPS into products like fish boxes, fishing floats and foam blocks. All machinery, except the steam boiler, and equipment can be fabricated by the entrepreneur and three supporting hands, preferably at least a mechanic and an electrician or an entrepreneur can make the easy ones on the floor and the rest be given out at small workshops.

### 11.2.1 The Products

-  Fish boxes – any standard size or sizes
-  Fishing floats – as shown earlier in this chapter
-  Foamed blocks – 3 ft. × 3 ft. × 1 ft. (thickness)

This plant could produce other items also, within the framework of the 2 m/t, like pipe insulation and hot/cold containers. EPS loose beads are also used for packaging.

### 11.2.2 Machinery and Equipment

1 No. steam boiler – single phase or other – 100 psi – diesel oil–water level indicator – safety valve.

(This steam source can be operated on rice hulls as fuel with some modification to the feed line.)

It is estimated that in the latter case, fuel costs will be reduced by 80% as compared to diesel. The main steam line from the boiler will be connected to a small accumulator from which the molding stations will draw the steam. This will help to prevent too much sudden draw of steam from the boiler.

1 No. steel/aluminum block mold with inner dimensions 36.25 in. × 36.25 in. × 12.5 in. This block mold will be made of steel on the outer sides and lined with perforated aluminum sheet in the insides, with very small perforations to prevent the EPS beads from going through. The design of the block mold should be such that there are at least two side openings (top and front) for easy removal of the molded foam block. This mold will also have inlets and outlet valves for steam and water placed at strategic places. Steam and temperature gauges fixed on the mold will help in easy and quality productions.

1 or 2 molding machines: These can be worked electrically or hydraulically with the base platen stationary and the top platen moving vertically up and down. The molds in two halves will be mounted on the two platens. Molds are generally made in aluminum with sufficient inlets through a steam jacket. Figure 11.4 shows a basic molding machine:

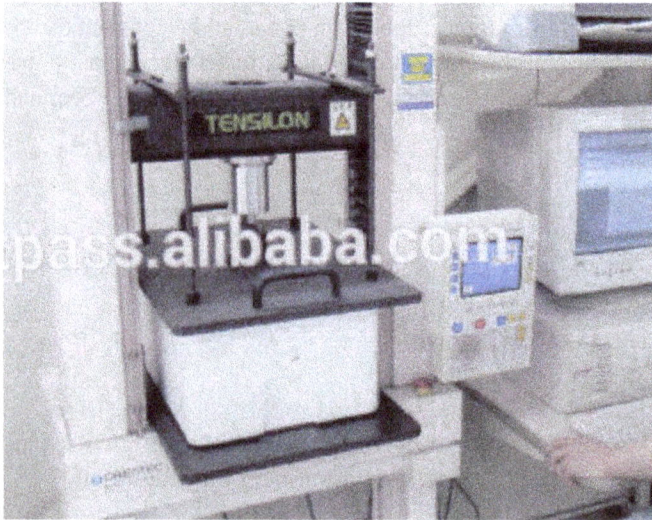

**Figure 11.4:** Shape molding machine.

1 No. preexpander can be easily constructed out of cheap sheets or even large steel drums. A large trough with a wire mesh bottom covered to prevent the foamed beads going through. This expander will have steam inlets from the bottom. Another open long tray is placed in front to accommodate the foamed beads and to remove any agglomerates that will occur.

Basic items required for constructing a hotwire cutting system are
- single phase 15 amp or 30 amp per controller 10–100 V;
- nickel/chrome 14 G and 16 G wire;
- a set of electrical clips (single wire) with spring attachments;
- two lengths of 150 cm × 1 cm aluminum channel;
- a 240 × 120 × 2.5 cm warp-free laminated board;
- a steel metal frame with legs to accommodate board at 45°.

Mount the board on the steel frame at 45° or horizontally (if table is to be motorized). The angled elevation will enable an operator to gravity feed the foam block. Cut long grooves on the aluminum channel to accommodate up and down movement of the hotwire/wires to be connected across the board. One may use a single wire or multiples which will increase the production output. Once the electrical wires have been connected, test the cutting wire until a "red glow" appears and allow it to cool. Since, during the cutting operation the wire will be on tension and breakages are possible. To avoid this, heat the wire to a red glow and beyond until a "gray" stage is reached, and allow it to cool before use. The height between the board surface and the cutting wire will determine the thickness of the foamed sheet to be cut. In multiple wire systems, it is possible to cut different thicknesses in one pass.

Molding system for fishing floats – the same shape molding machine can be used but it is advisable to have a second similar machine to accommodate either a 2-cavity or 4-cavity mold. Here, there must be a 25 cm hole through the float to allow the threading of a rope attached to fishing nets. If the plant is to make other products like hot/cold containers, small boxes or others, one may opt for a simple square or rectangular autoclave, where even multiple products can be made in one steaming/cooling cycle. In this operation, the autoclave is fully filled with steam which will enter each mold through the tine steam inlets provided, expand the preexpanded material further and mold the products.

Diesel oil generally comes in large 45 or 55 gallon drums. On the factory floor this can be fed into a smaller steel drum mounted on a steel frame on a height of about 8 feet which will have an outlet controlled by a valve, directly connected to the steam boiler fire chamber. The valve will control the flow to ensure a more or less "atomizing" effect to kindle a fire and then maintain the fire to heat the water in the boiler. The diesel from the main drum to the mounted source can be pumped using a hand pump or an electric pump and the top drum must have a window to indicate the level of diesel.

### 11.2.3 Raw Material

An entrepreneur would order the EPS beads in either small packs in paper bags or in larger packs in steel drums, small size beads for the fish boxes and fishing floats and a larger size bead for making large foamed blocks. These choices will depend on the final densities to be achieved and here a supplier's guidance would be valuable. One may opt for colored beads (more expensive) for hot/cold or other containers.

### 11.2.4 Production Method

Pump the diesel oil from the bottom drum to the mounted drum for gravity feeding to the boiler. Set up the steam boiler with the safety valve at 75psi and fill with water through a waterpump. The water gauge will indicate the desired level (should be more than half). Slowly gravity-feed the diesel oil to get a "cascade" effect inside the fire chamber. Ignite the oil through the "peep hole" and make sure there is a steady stream of diesel oil to ensure continuous fire. As the water gets heated, the steam gauge will show the rise in steam volume/pressure and when it reaches around 50 psi steam can be drawn for production through the steam accumulator, which will also have a steam valve to collect the condensing water.

An operator will introduce the EPS beads by spreading it evenly on the preexpander and open steam which will expand the beads. The steam should be unsaturated as otherwise it will scorch the material. This being only the preexpansion stage, the timing and size of beads will be important in relation to the final density. Generally, each batch will take only a few minutes and the operator will then open the preexpander and remove the expanded beads on to the long tray in front. He will then start a second batch and gently break up the foamed material separating the beads. If any agglomerates are found, he will gently break them up without crushing the beads. These foamed beads will then be packed into very large porous cloth bags and tagged with date and other data. This material can be used after about 24 h, after they have become dry.

These beads are then packed into the molds and steamed until they have formed products according to the shape of the molds. Cooling can be with water or it is vacuum dried. Shape molding is fairly straightforward but molding foamed blocks need a little more attention.

Fill the block mold with the expanded beads with a gap of about 50 cm from the top. This will allow the expanding foam to rise and occupy the full mold. Open steam and a timer and a steam gauge will guide the operator regarding the length of the steaming process. Too short a period will result in bad cohesion, while too long will make the block shrink and warp. Adequate cooling with water or vacuum is needed to prevent warping. After removing the block, it can be cut into sheets. Some processors may decide to trim all sides of the block before storage which will hasten the curing time by removing the water faster. However, cutting a "wet" block with a hotwire will be a little difficult.

### 11.2.5 Fabrication

The shape molded products do not need any fabrication but the fishing floats will need to have two inserts at either end to prevent chaffing due to the ropes being used in actual practice when being attached to the fishing nets.

Trim the foamed blocks and then cut the block into desired thicknesses. Every time a cut is made some material is lost due to the diameter of the cutting wire but in the case of nickel/chrome wire it is almost negligible. However, if a band saw is used, the loss would be bigger. Quality control random checks may include density checks and size checks using a wooden frame.

## 11.3 Manufacture of Polymeric Composite Resins

Over the past few years, as environmental concerns grew and polymer resin prices also increase, polymer processors have been looking for viable options. The advent of wood–polymer composites (WPCs) helped in this cause and as processor's confidence grew in using these composite resins, the plastics market expanded from using them in injection molding, extrusion and compression molding. The important aspect of using these composite resins other than their cost-effectiveness was that they could be processed on conventional machinery with different processing parameters.

For some time wood flour and chips were the leading reinforcing agents in the manufacture of composite resins but later diligent and constant research came up with rice hull powder or flour which could equally produce excellent composite resins with most common polymers. In practical use, it has been shown that these composite resins with rice hulls (PCRH) has even better properties than WPCs. See Figure 11.5, which shows extruded board from PCRH resin.

Figure 11.6 shows a flow diagram of the concept to final application of polymeric composite resin products as designed by the author.

### 11.3.1 Technology in Brief

Technology for manufacturing these PCRH has come a long way. The possibilities with these versatile resins are vast and are proving to be suitable for even high-tech applications in automobiles and space travel. The basic technology is that polymer resins such as polyethylene (PE, HDPE), polypropylene (PP) and polyvinyl chloride (PVC) are used as the matrix and mixed with rice hull fiber but preferably in the form of powder or flour and additives, depending on the end application. Some applications will need the addition of a combination of additives. The polymer matrix can be virgin resin, recycled material or a combination of both. Additives can take the form of ultraviolet (UV) stabilizers, coupling agents, heat stabilizers, lubricants,

**Figure 11.5:** Extruded board from "PCRH resin" (reproduced with permission from Harden Industries Ltd., China).

antibacterial agents and so on. Colorants and other surface agents are added if it is necessary to obtain colored or aesthetically pleasing finishes. Some can produce an identical wood-grain effect.

Rice hulls should be sieved and dried thoroughly to remove moisture before being compounded with polymers using heat. Additives are also added before compounding and introduced into an extruder having a screen pack to remove any further foreign matter if any and the multicavity die will extrude continuous strands going through a cooling process and then through a pelletizing unit which will cut these strands into predetermined pellet sizes. These pellets will be opaque to slightly yellowish in color or can be fully colored if colorants are used. These pellets can now be used for extrusion, injection molding and compression molding.

### 11.3.2 Raw Materials

As already mentioned, the basic raw material are a polymer matrix, which can also be a combination of polymers, rice hulls powder or flour, additives and colorants (optional). Each formulation will differ depending on the end application. Surface additive agents can produce either a matt or gloss finish. PCRH products tend to be hard or heavy and if lighter products are desired, a blowing agent can be incorporated into the mix. For example, hard or heavy products may be ideal for flooring and building construction, whereas lighter aesthetically pleasing products will be ideal for automotive, indoor applications or others.

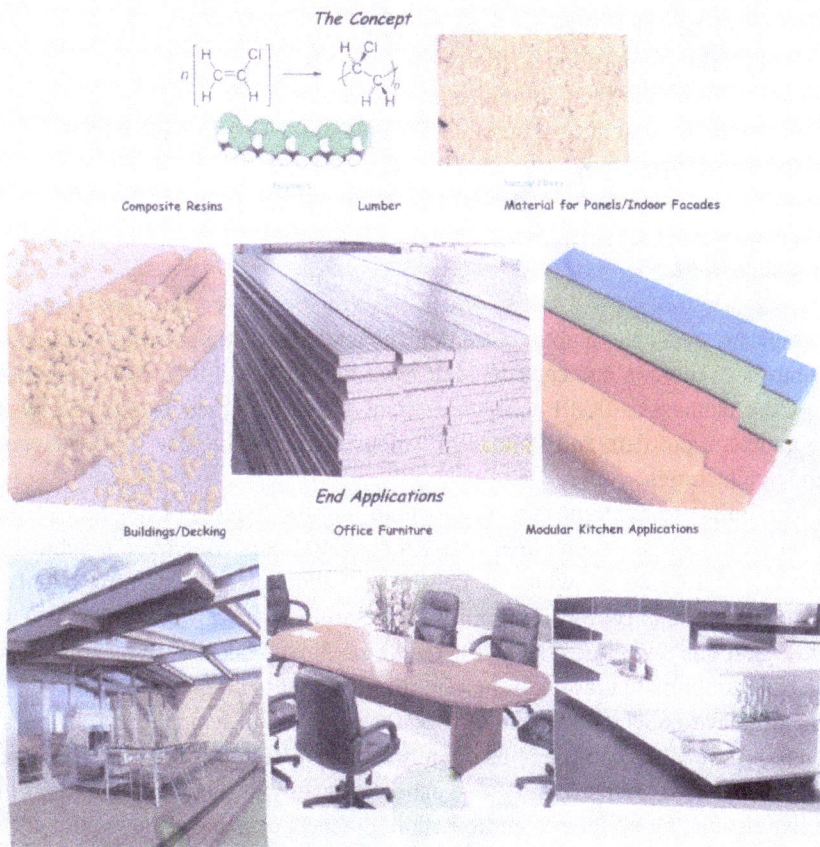

Project Consultant: Chris Defonseka Suite 814-11753 Sheppard East Toronto Ontario Canada M1B 5M3

Ph: 1 416 282 0002 mobile: 647 629 0799 e: defonsekachris@rogers.com

## Polymer Composites Lumber with Rice Hulls

- *A pioneering project for Sri Lanka* -

*The Concept*

Composite Resins          Lumber          Material for Panels/Indoor Facades

*End Applications*

Buildings/Decking          Office Furniture          Modular Kitchen Applications

**Figure 11.6:** Flow diagram for extruded PCRH.

### 11.3.3 Polymers

In general, thermoplastic polymers like PE, PP, PVC, acrylonitrile butadiene styrene and PS are the easiest to work with. Combinations are also feasible. They could be in pellet or powder form free of atmospheric moisture. Some formulations may use virgin or recycled material or a combination. If different colored materials are recycled together, it could be difficult to obtain a satisfactory final color from the resulting product. If mixtures of polymers are used as the matrices, it is important to use compatible polymers with similar melt-flow properties. To aid these mixtures coupling agents could be used. Polymers are generally available in 25 kg bags or 400 kg "Gaylords." The proportions of the polymer matrix to the reinforcing agent/agents will vary according to the end application but a base guide would be around 40:60. Higher rice hulls contents are possible. To make these final products lighter, a compatible blowing agent may be used.

### 11.3.4 Rice Hulls

Rice hulls are available in abundance in many countries. Until recently, this gift of nature has been more or less considered as a waste with very little use. The possibilities with these biomass as PCRH are both exciting and challenging. These are now available as fibers, powders and flour from various suppliers for easy use for composite resin manufacturers.

Rice hulls are different to wood fibers in that the lengths of rice hull fibers are only around 0.3 mm as compared to 6–7 mm wood fibers. Moreover, rice hulls have a content of 20% silica, which is helpful as stiffening/strengthening agents and moisture barriers in composite products. Rice hulls in the raw form may have loose surface add-ons and "fines" which should be removed before use by some method such as a tumbling operation or on a fluidized bed with air.

Tests have shown that to maximize the strength of a final product, it is not enough just only to clean the hulls but any dust or foreign matter should be removed from the "cleaned" hulls. This can be achieved by any of the commonly used methods like sieving, air cyclone or air floatation. Practical tests have shown that grinding the rice hulls in a hammer mill followed by sieving will produce good "clean" material. The ideal would be to finally have the rice hulls in fine powder form for easy blending with the polymer matrices.

Being an agricultural waste, rice hulls will have an inherent moisture problem, especially, if left out in the open. Due to the sheer large volumes produced, it is very difficult to protect them from the elements. Buyers of the raw rice hulls from the mills for conversion will have to contend with moisture and foreign matter which will invariably be present to some extent. The preferred moisture content of the rice hull powders should be around 1.0% or less and certainly not more than 2.0% to prevent swelling and blisters at the time of extrusion.

### 11.3.5 Additives

Additives are an essential part of making polymeric composite resins, some as processing aids, some to enhance properties and others to achieve aesthetically pleasing finishes. In the manufacture of PCRH resins, one must remember that due to the presence of silica in the rice hulls, the mixture will be abrasive, generating extra heat in the extruder chamber and also may speed screw and barrel wear. To counter this action, processors will use an adequate lubricating additive and extruder manufacturers have also come up with advanced steel materials to counter this wear and tear.

Some of the basic additives used are
- lubricants;
- stabilizers;
- anti-UV agents;
- flame retardants;
- coupling agents;
- antifungal and microbial agents;
- blowing agents;
- surface finish agents;
- colorants.

### 11.3.6 Processing Machinery and Equipment

The complete process will consist of five main operations: sieving, drying to remove moisture, particle size reduction, compounding and extrusion into strands and pelletizing. The first three operations can be done separately but a processor may opt for a complete in-line compounding, extrusion and pelletizing system. The mixture of polymer/polymers, reinforcing agent/agents and additives is put into an extruder, the extrusion head having a profiled die of choice. The strands coming out of the multidie will be cooled as they emerge from the die and taken up downstream and a pelletizing unit will cut the strands into desired size of lengths. These will be stored in silos and then fed into a larger extrusion system and extruded into profiled sections or flat sheet, depending on the production needs. If large volumes of composite boards are to be produced, the pelletizing stage can be eliminated and from the compounding stage, the second larger extruder can extrude flat boards using a T-die in continuous form. A vertical cutting system mounted on the extrusion line downstream will cut the solid boards into desired length. The T-die will determine the width and thickness of the board to be extruded. As the extrudate emerges from the die, it will expand as it is exposed to the atmosphere and will go through a size calibrating unit before being cooled.

These PCRH boards will be an ideal substitute for natural wood and will also have superior properties. The author's book *Polymeric Composites with Rice Hulls* (ISBN978-3-11-064320-6 – De Gruyter) gives a detailed presentation. In the composite world, the word *WPC* refers to polymer composites with wood but PCRH refers to composites with rice hulls. When WPCs appeared on the market, they showed a great potential but when researchers realized that it was possible to make polymeric composites with other natural fibers, especially biomass, some of the world's top machinery manufacturers such as *Reifenhauser, Cincinnati Milacron, Maplan Corporation* and *Davis Standard* began to develop specially designed extrusion system to meet these new challenges. With rice hulls, the real challenge was to overcome the extra wear of the screws due to excess heat generation due to the presence of silica in the rice hulls. These suppliers of extrusion systems not only supply extrusion lines but work closely with processors to ensure quality productions in mutually profitable ventures. Their in-house research laboratories are a great asset to processors.

Although PCRH composites stand out above others, when using other natural fibers, processors have to adhere to certain basics to ensure ease of processing and quality products. Recommended specifications for the use of fibers to achieve the best material properties and highest production output rates are

- Fiber size 0.1–1.0 mm, ideally 350–500 µm, equal to mesh numbers 35, 40 or 45.
- Moisture content: <12% for direct extrusion or <3% compounded pelletized extrusion.
- Fiber type: generally a wide range of fibers are possible but soft ones are the easiest to process.

PCRH products have a wide range of end applications. In North America and Europe, outdoor applications took precedence but following the practices of Asian countries and China, newer applications have been taking place, spreading into industries like the automobile and building construction industries. The versatility of PCRH products has spread rapidly and now includes sectors such as indoor and furniture industries. The toughness, warp-free and moisture resistant properties of these materials has made these PCRH lumber ideal for railway sleepers among many other similar applications. Great surface finishes such as imitation wood veneers in matt or gloss finishes makes them ideal for flooring, indoor interior applications, while the laminated boards shown in Figure 11.5 can be used for kitchen cupboards and others with minimal costs being able to be fabricated with conventional wood-working tools.

## 11.4 Manufacture of PUR Flexible Foam Mattresses

A flexible PUR foam is a soft, open cell and pliable material belonging to the *thermo-setting group of plastics* and is made by a combination of several chemicals. The main components are polyols as the base, toluene diisocyanate (TDI) or di-phenylmethane di-isocyanate, water and methylene chloride (blowing agents), amine catalysts and tin catalysts.

The incorporation of additives will give the foam special properties such as color, UV protection, high resilience, fire retardation and antifungal protection as desired. These foams can be made in different densities and qualities to meet the needs of the final end applications. A manufacturer of these foams must have a good knowledge of the technology involved or at least have an in-house chemist or a trained person to be able to formulate effectively. Polyols can be based on polyether or polyesters and although both will produce open cell foams, the latter, due to its smaller cell structure will have tighter air flow and it is less flexible. Polyether polyols will produce softer foams and the preferred material if comfort is the primary consideration.

Eco-polyols, for example, made from soya will also produce open cells but will give lower yields and the inherent slight odor has to be masked. Foams made from all three types of polyols are widely used for making PUR foams in very large volumes all over the world.

Viscoelastic foams (memory foams) are also based on polyols and can be classified under flexible PUR foams. They are produced more or less by the same chemicals, except that special grades of polyols are used.

### 11.4.1 The Raw Materials

Here the author presents a description of the raw materials in brief for the benefit of the reader but the more interesting aspects are the formulations and production methods, which are presented later.

**Polyols:** Polyether polyols are long-chain alcohols that are made by polymerizing common hydrocarbon oxides. This results in linkages that connect the hydrocarbon portions of the chain and hydroxyl functional groups at the end of the chain. Triols with molecular weights from 3,000 to 4,000 made by copolymerizing a mixture of propylene oxide and ethylene oxide with glycerine as the initiator are often used to produce conventional flexible slabstock from which foam mattresses are made.

Graft polyether polyols contain copolymerized styrene and acronitriles. They are designed to maximize loadbearing properties in PUR foams. They contain solids in the range of 10–45% and sometimes used in the production of slabstock (foam blocks).

Biopolyols started emerging in the market due to global environmental concerns. Biopolyols are derived from vegetable oils such as soya, canola and peanut oil and are made by different methods. These polyols are specially made to make PURs. All are clear liquids, ranging from colorless to slight yellow. Their viscosities also vary and are usually a function of the molecular weight and the average number of hydroxyl groups per molecule. Inherent odors will differ from polyol to polyol but most bio-polyols are similar to the original parent vegetable oil and are prone to becoming rancid with time.

**Isocyanates:** The most common isocyanate used in the manufacture of flexible PUR foams is TDI. MDI can also be used. TDI is a low-cost, high quality product that allows manufacturers to produce many different types of PUR foams with a wide range of physical properties. There are also different types of TDI and one could also use blends if necessary.

Isocyanates are liquids and available in drums, totes or tankers for the very large volume producers. The estimated shelf life is about 6 months. Due to its high flammability and corrosive properties, suitable precautions must be taken at all times and standard precautionary practices must be observed.

**Catalysts:** Catalysts are needed in the production of PUR foams because two major reactions take place. In the polymerization or gelling reaction, the polyol will react to form PUR. In the gas-producing reaction, the water reacts with the isocyanate to form polyuria and carbon dioxide. These reactions occur at different rates, with both reactions mainly dependent on temperature, catalyst types and catalyst levels. To produce high quality foams it is essential to control both these reactions effectively and thus, the need for catalysts.

Tin catalysts like stannous octoate are widely used to catalyze and control the polymerization reaction. Insufficient catalyst will lead to foam splits or polymer collapse. Excessive catalyst levels will cause closed cells and shrinkages. In contrast to tin catalysts, amine catalysts catalyze the gas-producing reaction. Any residual catalyst will escape from the finished foam or incorporated into the polymer structure.

**Blowing agents:** Water and methylene chloride are the blowing agents, with water being the primary blowing agent. The latter functions as an auxiliary blowing agent. Others can also be used but some are banned due to environmental issues. Water is a key component in any PUR formulation but has a maximum threshold limit around 5.0 pbw. of polyol. Excess use will be prone to fire hazards. Water reacts with isocyanate to form compounds which remain in the foam and also carbon dioxide, which acts as a blowing agent. If required, auxiliary carbon dioxide can also be added as a liquid to augment the blowing portion of the foaming process.

**Surfactants:** Flexible PUR foams are made with nonionic, silicone-based surfactants. Different grades of surfactants are available to a foam manufacturer to meet specific needs. Silicone surfactants carry out the following five main functions:
- Reduction of surface tension.
- Provision of film resilience known as "self-curing in the bubbles."
- Control of cell size to form homogenous fine cells.
- Provides bubble breakage at full rise preventing shrinkage.
- Prevents deforming effect of any solids added to the reacting system.

Of all these factors, the most important is the stabilization of the cell walls. Surfactants prevent the coalescence of rapidly forming cells, until they have attained sufficient strength through polymerization to become self-supporting.

**Methylene chloride:** It is a liquid with a low boiling point. It is used in foam formulations to obtain lower densities and extra softness, generally, not possible with the primary blowing agent – water. Methylene chloride absorbs the heat from the exothermic reaction of foaming, vaporizing and providing additional gases helping the expanding foam to a lower density. These liquids are also used for cleaning and flushing.

Methylene chloride acts as an auxiliary blowing agent in a PUR foam formulation by complimenting the blowing effects of carbon dioxide generated from the water/isocyanate reaction. These liquids are available in steel or plastic drums and standard precautionary measures must be observed.

**Additives:** They are components incorporated into the PUR formulation to achieve specific properties for particular end uses. They do not interfere with the basic chemistry of the foaming reactions. Some of the common additives used are
- colorants;
- fillers;
- flame retardants;
- antioxidants;
- cell openers;
- plasticizers;
- antibacterial agents;
- antistatic agents;
- UV stabilizers;
- heat stabilizers;
- foam hardeners;
- cross-linkers;
- compatibilizers.

**Colorants:** Foams made from general formulations systems are colorless or white. Pigments are added to the polyol to color the foam blocks for easy identification, mainly for identification different densities. This is a great help for large volume foam producers where many different densities and foams with special properties are produced. Yellow is the preferred color for standard productions to counter UV action, while any colored foams helps in the aesthetic value. Liquid colorants are preferred but if powders or others are used, it must be thoroughly mixed in the polyol for homogenous foam color. Typical problems that may be encountered by using pigments are
- foam instability;
- foam scorch;
- color migration;

- limited usable color range;
- abrasive action on pumps and mixers.

**Fillers:** They are generally fine particles of inert inorganic compounds and are added to foam formulations to increase density, load-bearing and sound absorption with the added advantage of cost reduction. The use of fillers may affect certain physical properties of foams.

Of all the wide range of fillers available, only the inorganic calcium carbonate fillers are most widely used in foam productions. If fillers are used with diligence, substantial cost reductions can be achieved. An important aspect of fillers is that they should be very dry or at least a formulator must know of any moisture content in them and preferably as a percentage, to enable necessary adjustments in the water content so as not to exceed the allowable threshold.

**Fire retardants:** Low-density, open-celled flexible foams have a large surface area and high permeability to air. PUR foams will burn given a sufficient source of ignition and oxygen. Flame retardants of choice are often added to reduce this flammability. The choice and grade of flame retardant to be used will depend on the scope and end application of the foam. This will also depend on local codes for different end applications.

Some aspects of flammability to be considered when choosing a grade are intended application, initial ignitability, rate of burning and smoke evolution. Many grades are available from chemical product suppliers and the most common ones used are chlorinated phosphate esters, chlorinated paraffins and melamine powders. Manufacturers can be guided by the suppliers on which grade or grades to be used. Laboratory tests can also be carried out, if necessary.

**Ultraviolet stabilizers:** PUR general foams are colored yellow which counters UV affects to some extent when exposed to direct sunlight. All PUR foams based on aromatic isocyanates turn a brownish to dark yellow color with aging along with prolonged exposure to sunlight. This "yellowing" is not a problem for most foam applications. Protection agents such as zinc dibutyl thiocarbamate, hindered amines and phosphites can be used to improve the light stability of PUR foams.

The properties of all water-blown conventional polyether PUR foams are controlled by the formulations and processing conditions, more so than individual component properties. Very small quantities of catalysts and surfactants are used in a formulation for good quality high airflow foams for furniture, bedding and carpet padding. Therefore, only the major effects of water, isocyanate and polymer solids need close attention. The term "polymer solids" is used interchangeably with "parts of polymer polyol" although these two differ by a constant factor. Processing aspects includes those that affect foam breathability, structure and cell count. For manufactured foams, the important factors are density and indentation force deflection (IFD). The general rule is that as the water content in a formulation increases, the density and IFD will decrease, with higher contents of isocyanate giving softer foams with lower densities.

Flexible PUR foam manufacturing technology is quite complex. Thorough knowledge and experience is needed to formulate correctly with atmospheric conditions also playing a part. Warmer processing conditions will give good quality foams. Generally, the primary source of knowledge is with the manufacturer or a member of the staff who has experience or has been specially trained. The secondary source is the supplier of the raw materials, who can provide a wealth of information and datasheets on all products purchased. In foam manufacturing, it is always advisable to have an in-house laboratory to carry out tests of different formulations to ensure quality foams and also to minimize waste.

## 11.4.2 Machinery and Equipment

There are three basic manufacturing systems: manual operations (small volumes), batch production (single blocks) called the *intermittent process* (medium size volumes) and the *continuous process* (continuous foaming – very large volumes). Only the first two production methods are presented and any reader who is interested in obtaining more information on the continuous process may refer to the author's book *Practical Guide to Flexible Polyurethane Foams* (ISBN 978-1-84735-974-2), where this process is presented in full detail.

**Manual operation:** Where small blocks of foam can be produced, for example, cushions for furniture. All needed machinery and equipment can be easily fabricated on the production floor or purchased from the market. Molds can be made of wood, an electric drill as a mixer, a small weighing machine, a hotwire cutting system and safety wear would be the basic equipment needed. This process is presented in detail later.

**Intermittent process:** Where single large blocks can be made one at a time. The molding/mixing unit is stationary, while the molds on wheels are brought in for each pour and wheeled out and the next mold is brought into place. With three or four molds, fairly large volumes of PUR foam blocks can be made. All chemicals are connected to the mixing head and weighing is carried out electronically. Basically, a large mixing vessel is lowered onto the inside mold base and after mixing of the components, the cylinder automatically lifts up, allowing the rising foam to fill the mold. Figure 11.7 shows a basic machine.

**Continuous process:** This process is used by manufacturers who make very large volumes of foam. The main components are purchased in bulk, while the additives are in small ecopacks. All components are connected directly to the main mixing head and the amounts are electronically weighed and mixed in a central mixing head. Once set, the mixing head delivers continuous foam onto a slowly moving conveying trough with the foam rising to a preset height while moving

**Figure 11.7:** Basic machine (reproduced with permission from Modern Enterprises Ltd., India).

forward and is cured by the time it reaches a vertical cutting system. This system will cut the foam into desired lengths. Figure 11.8 shows a continuous foaming line.

**Figure 11.8:** A continuous foaming system (reproduced with permission from AS Enterprises Ltd., India).

### 11.4.3 Calculating Density

In PUR foam manufacturing, density factor becomes very important. In formulating, one of the main criteria is to formulate to achieve a predetermined density:

$M = V \times .D$ where $M$ in kg, $V$ represents volume (m$^3$) and $D$ represents density. Therefore, density in kg/m$^3$.

### 11.4.4  Calculating Indentation Force Deflection

The IFD as a number is an important marketing tool. This is really an indication of the supporting factor for a PUR foam. A value below 2.0 is considered below par, while IFD values above 2.0 are considered acceptable. These apply especially to bedding and furniture industries.

IFD is a ratio of compression of a selected foam sample at 25% and 65%. The sample used for testing will be $60 \times 60 \times 10$ cm being compressed by an indenter plate of an electro-mechanical device. The ratio of the values obtained is 65%:25% and is known as the IFD.

## 11.5  PUR Flexible Foam Manufacturing Methods

Here, the author presents a manual operation and an intermittent process, keeping in mind the benefits for an entrepreneur or a small volume molder. These presentations are based on actual productions done successfully by the author in his factory in Sri Lanka but the reader must accept them as guidelines only as conditions and components will be different when carried out in different locations.

### 11.5.1  Manual Operation

Objective: To produce foam cushions size $50 \times 50 \times 10$ cm or any other thickness. Ideal as a start for an entrepreneur with limited resources or as an in-house operation for a furniture manufacturer.

Basic needs would be as follows:
(a)  A building with a floor area of about 200 m$^2$ with single-phase or three-phase power and standard water supply. The building will have easy access to shipping and receiving and adequate ventilation. Space for storing raw materials (corrosive and flammable) and for storing cured blocks (24 h) before fabrication.
(b)  Machinery: A single-phase handheld standard electric drill with adjustable speeds of 900–1,200 rpm with a forward/reverse function to be the mixing device. A steel rod is of approximately 1.25 cm in diameter, 75.0 cm in length with a 10 cm diameter disk with four flanges firmly attached to one end of the rod. The flanges should be facing down when the rod is connected to the drill. This arrangement will act as a mixer for the chemical components.
(c)  Basic requirements for equipments would be as follows:
  –  0–2 kg and 0–10 kg electronic weighing machines;
  –  4–5 large plastic buckets;
  –  glass beakers and a few test tubes;
  –  glass rods and stirrers;

- magnifying lens and steel rulers;
- two drum stirrers;
- two manual or electric pumps or;
- two horizontal drum stands;
- two valves with taps for drums;
- miscellaneous tools.

(d) Raw materials:
- polyether polyol in drums;
- TDI (isocyanate) in drums;
- methylene chloride (optional);
- water source;
- Desmorapid OS20 or equivalent;
- Desmorapid PS207 or equivalent;
- Desmorapid SO or equivalent;.

Since space is limited it is advisable to purchase the polyol and isocyanate in small quantities like one metric ton at a time. Just-in-time purchasing would be ideal and economical, if it is feasible. For easy identification a supplier will pack these items in two different colors, such as blue and red or green and red. All others being used in very small quantities should be purchased in standard ecopacks. Bulk purchases will be cheaper and will not be practical for a small molder.

For all materials, especially the polyol and TDI, should be stored at the correct temperatures, as recommended by the supplier, which can be around 25 °C or less. It is preferable to have the contents in these two drums stirred well before use. These drums can be mounted on the drum trolleys and by attaching a drum faucet (tap), the liquids can be easily drawn out. The smaller containers with other components will not pose a problem and the required amounts can be easily weighed. Since these additives are small quantities, some of them can be measured into a test tube which has a permanent "level mark," thus saving weighing time for each batch. At all times, operators must wear safety wear and if any of the chemicals comes in contact with the skin, it must be washed with water immediately.

## 11.5.2 Molds

Can be easily made from wood or similar cheap warp-free material. The mold base must have grooves to accommodate the vertical sides of the mold. For making eight foam cushions of dimensions $50 \times 50 \times 10$ cm each, the inside dimensions of the mold when assembled should be $51 \times 51 \times 88$ cm (height). For easy removal of molded foam block, it is best if all four sides can be removed. The mold construction can be either square or rectangular depending on the end requirement but for this exercise the mold should be square.

The mold sides will be sprayed with a release agent like silicone or other or lined with very thin polythene film for easy removal and also to prevent the hot molded foam adhering onto the sides which will increase the thickness of the trimming. Also needed will be a square piece of very light wood – $50 \times 50$ cm with a handle on one side of a surface to be used as a "floating lid" to prevent a "meniscus" forming on the surface of the rising foam. Molds can be on wheels for easy movement and availability of more than one mold will increase production volumes.

### 11.5.3 Cutting and Fabrication

Foamed blocks have to be "cured" for at least 24 h before they can be taken for cutting. Best to use the *first in, first out* (FIFO) system as otherwise semicured foam blocks will result in great waste. Foam cutting can be done with a hotwire system or a band saw machine. While the former will give very smooth cuts, the latter will result in slightly more "waste" because of the thickness of the cutting blade. For example, if a hotwire is used to cut a 50 cm thick foam, one could get 10 pieces of 5 cm thick, while a band saw will probably give extra 9 pieces.

From the foam block produced it is possible to cut thin sheets, slabs, cushions, wedges and so on by angling the hotwire. If a grooved hotwire is used, patterned surfaces could be cut. A foam block taken for cutting will undergo trimming on all sides to get rid of the uneven surface and these will have to be recycled to minimize waste. A shredder could produce small pieces for packaging or making pillows. Large volume foam manufacturers have their own in-house recycling machines which converts foam wastes into compressed large blocks from which slabs are cut for mattress bases or sheets for carpet underlay. Generally, foam wastes are high, could be around 15% or even higher, with recycling a must.

### 11.5.4 Production Method

Exercise: To make a foam block of density = 29 kg/m³ and to cut 8 cushions of dimensions $50 \times 50 \times 10$ cm. Assuming the required block size would be $51 \times 51 \times 83$ cm, the required pour calculation would be as follows:

Applying, $M = V \times D$

$V = 51 \times 51 \times 83$ cm
  $= 215{,}883$ cm³
  $= 0.216$ m³

Therefore, $M = 0.216 \times 29$
      $= 6.26$ kg + 1% waste = 6.32 kg/pour
Note: Waste factor/gas allowance can vary from 1% to 5%.

**Formulation and calculations as recommended by the author are as follows:**

| Component | Quantity (kg) | Calculation | Batch quantity (kg) |
|---|---|---|---|
| Polyol | 30.000 | 30/56.1 × 6.32 | 3.38 |
| TDI | 21.350 | 21.35/56.1 × 6.32 | 2.41 |
| Water | 1.825 | 1.825/56.1 × 6.32 | 0.20 |
| Methylene chloride | 2.210 | 2.21/56.1 × 6.32 | 0.25 |
| OS 20 | 0.560 | 0.560/56.1 × 6.32 | 0.06 |
| PS 207 | 0.059 | 0.059/56.1 × 6.32 | 0.01 |
| SO | 0.061 | 0.061/56.1 × 6.32 | 0.01 |
| Total | 56.065 | | 6.32 |

From above, each batch pour will be 6.32 kg. Foaming method is as follows:

Set up the mold and line with thin polythene sheet or wax the sides or spray a release agent. Ensure the mold will not leak any material which when poured will be in a liquid state. Weigh all components in different containers. Use a large plastic bucket for the polyol. Add a small quantity of colorant (yellow or other) into the polyol and mix well for 60 s. Mix water, PS207 and the OS20 together and add to polyol and mix for 15 s. Add methylene chloride and SO to the mix and mix for a further 15 s. Add TDI to this mix and mix for 3–4 s and pour mix into the mold, still in liquid state and before it starts to "cream." If this happens, the whole batch will be a reject.

The chemical reaction will start with the mix becoming cream in color and the foam begins to rise inside the mold. When the foam has risen about one-third the height of the mold, place the "floating lid" gently on top of the foam. The foam will rise with the lid and when it has stopped, remove the lid. The next batch can be initiated now with another mold. After about 10 min when the molded foam block top is not tacky, demolding can take place. Move the foam blocks to a well-ventilated holding/curing area and store them at least one foot apart. This will prevent possible fire hazards from gases being given out from the foam blocks due to the continuing exothermic (heat giving) reactions. These blocks can be taken for cutting after 24 h.

## 11.6 The Intermittent Process to Make Large Foam Blocks

This section shows how an entrepreneur or a small foam volume manufacturer can make large foam blocks using an intermittent (discontinuous) method. The production of these large foam blocks necessitates more sophisticated machinery than for making foam cushions. Many choices are available from different machinery manufacturers but the foaming principle will remain the same, while the methodology may vary slightly. These machines are available in manual or semi-auto foaming operations. Products that can be made from these blocks are mattresses, cushions, sheets, slabs and contoured shapes.

These machines will be connected to all component sources and will have automatic control with component feed systems in sequence to a central mixing head, with timers for mixing, speed adjustments, lowering of mixing cylinder to mold flow, automatic time set for opening of bottom lid to allow the mixed liquid mixture to flow out and spread evenly on the mold floor and also automatic lifting of mixing cylinder. These machines will give fast molding cycles, with availability of multiple molds on wheels and could produce 30–40 blocks per 8 h. An effective and economical operation may be based on two days of foaming per week with continuous cutting and fabrication. Foam volumes, of course, will depend on marketable volumes plus a reserve.

Basic production data:
- foaming density – 16–32 kg/m$^3$;
- foam block size – $2 \times 1 \times 1$ m or $2 \times 2 \times 1$ m;
- estimated output – up to 40 blocks per 8 h.

Basic equipment needs
- machine space – $4 \times 8$ m;
- air compressor – 2 HP;
- electronic weighing scale 0–2 kg and 0–300 kg;
- electrical power – 3 phase;
- trolley to move foamed blocks to curing/holding area.

Note: Size of molds can vary but the volume of a single pour will depend on the shot capacity of the machine.

Basic foam cutting machines
- 1 No. circular cutting machine (carousel type) optional;
- 1 No. vertical cutting machine;
- 1 No. horizontal cutting machine;
- 1 No. foam shredding machine.

There are many other types of fabricating machines among which the peeling machine (for making continuous foam sheets), the contour cutting machine (different shapes) and the carousel machine (cutting large volume mattresses) may also be of interest to a PUR foam manufacturer. Since band saws are used in these machines for cutting, an allowance may be made for the thickness of the cutting blades when cutting blocks of foam. For example, a cutter may not be able to get two mattresses of 10 cm thickness each from a foam block of 20 cm.

Note: Unlike other plastics, foams are not easy to shred due to its softness and flexibility. In general, a 10 HP high-speed shredder with special blades is needed to obtain at least 10 mm size pieces for rebonding. Some of the large size "trims" from the block sides does not have to be shredded as they can be sold for packaging. Another option of shredded foam pieces is for making pillows or even for making PU adhesives.

### 11.6.1 Production Example – Intermittent Process

Example: To produce a large foam block to cut 12 twin-size mattresses, each $72 \times 48 \times 4$ inches thick of density 29 kg/m$^3$.
Required foam block size = $73 \times 49 \times 50$ inches (height)
Note: Make an allowance of 0.5 inch for side trims and 1.0 inch for top and bottom.

Applying, $M = V \times D$

$\qquad$ Volume = 183 cm $\times$ 123 cm $\times$ 125 cm = 2.8136 m$^3$

$\qquad$ Then, mass = 2.8136 $\times$ 29 = 81.60 kg

Therefore, each foam pour will be = 81.60 + 1% waste = 82.41 kg

Then, using formula,

| Component | Quantity (kg) | Required quantity (kg) |
| --- | --- | --- |
| Polyol | 30.000 | 44.10 |
| TDI | 21.350 | 31.36 |
| Water | 1.825 | 2.70 |
| MC | 2.210 | 3.25 |
| OS 20 | 0.560 | 0.82 |
| PS 207 | 0.059 | 0.09 |
| SO | 0.061 | 0.09 |
| Total | 56.065 | 82.41 |

**Foaming method**: Prepare the mold and wheel it in under the machine mixing cylinder and lower the cylinder onto the base of the mold. Meter the polyol into the mixing cylinder, adding any color desired and mix for 60 s. Mix the catalysts and water and add to the polyol and mix for 15 s. Add methylene chloride and SO to the mix and mix for 15 s. Add the TDI to the mix and mix for 4 s. Since all these timings are preset and work electronically, they will activate automatically. The mixing cylinder will rise opening the bottom lid, releasing the liquid mix onto the mold floor and spread evenly. Creaming will take place as the chemical reactions will produce foam and the whole mass will rise slowly. If a flat top is desired, place a very light "floating lid" on top surface after the foam has risen about one-third of the height of the mold.

When the foam has ceased to rise, allow a few seconds for it to settle and then wheel out the mold on to a holding area and remove the lid. When the cylinder "washing mode" is finished, another mold can be wheeled in and another cycle started. Figure 11.9 shows some large foam blocks.

**Figure 11.9:** Large molded foam blocks. Photo from author's factory in Sri Lanka.

Allow the foamed block in the mold to stand for about 10 min and when the top surface is tack-free, demolding can take place. To minimize foam loss due to a possible "meniscus" on the top surface, stand the foam block "upside down" as shown in photo earlier. This foam being still warm will flatten the top surface. This should be done soon after demolding and before these blocks are moved to the final curing/holding area. Store the foam blocks with space between them. This area should have good ventilation and preferably an exhaust system. All blocks should be marked with at least date and density for identification purposes and for using the FIFO system to ensure only fully cured blocks are taken in sequence for cutting/fabrication. The introduction of a quality control system for this section will be of great value. As mentioned earlier, the two basic factors for mattresses are density and IFD, which should be ≥2.0.

If resources permit, a foam molder can have an in-house rebonding machine for shredded foam wastes, which should be substantial due to the production of large blocks. Basic types of rebonding machines are steam/compression or adhesive/compression where large thick blocks of foam are produced to be cut and used for mattresses bases, carpet underlay and so on. Alternatively, a foam molder could have the wastes shredded and rebonded byoutside sources or sells all the wastes as it is. Some of the small pieces of shredded wastes may be used for making pillows or also sold to PUR adhesive manufactures.

## Bibliography

[1]   Diversifoam *Products*-Technical Information. www.diversifoam.com/xeps.htm.
[2]   BASF – Expandable Polystyrene- Technical Leaflet S-2. www.basf.com.
[3]   Omnexus-"Expanded Polystyrene (EPS). https://omnexus.specialchem.com
[4]   Molded Products, "The Molded Polystyrene Manufacturing Process", www.achfoam.com.
[5]   Canadian Plastics Magazine, Feb 2019/Nov 2018/ Sept 2018, www.canplastics.com.
[6]   Defonseka, Chris. "Practical Guide to Flexible Polyurethane Foams", Smithers Rapra, 2013
[7]   Defonseka, Chris, "Polymeric Composites with Rice Hulls", De Gruyter, 2019.
[8]   Modern Enterprises Ltd. – India, www.foam-machinery.com.

# Appendix A

The following table shows some randomly selected suppliers of biomass fillers and stiffening agents. Since most are manufacturers, customized products can be purchased. Best sources for finding suitable suppliers for additives, dyes, pigments and others are these biomass suppliers who can also recommend the best products to use.

| Supplier | Country | Products |
| --- | --- | --- |
| Composite Materials Co. Inc. | USA | Wood flour, walnut shell flour, rice hull flour, sisal and corn cob flour |
| Hammond Roto-Finish | USA | Rice hull flour |
| Mid-Link International Co. Ltd. European office/Shanghai, China | Germany | Rice hull ash |
| Silicon India | India | Rice hull ash, powder and pellets |
| NK Enterprises | India | Rice hull ash |
| ADF Asset & Investments | UK | Wheat hulls |
| Agrilectric Research | USA | Rice hull ash |
| Tianjin Glory Tang Co. Ltd. | China | Bamboo fiber |
| Siddhi Vinayak Enterprises | India | Bamboo fiber |
| Zenco Global Enterprises | Malaysia | Soybean flour and wheat hulls |
| M.M. Chemical India | India | Composite polymer powders, HDPE, LDPE, PP, EVA and customized powders |
| Kanju Industrial Ltd. | China | Graphite powder for polymers |
| Rice Hulls Specialty Inc. | USA | Rice hulls and rice hull powder |

This table, compiled by the author, shows some key suppliers but many more are available, covering a wider range of products.

https://doi.org/10.1515/9783110656152-012

# Appendix B

The following table shows some suppliers of machinery whose processing systems are suitable for processing polymeric materials into final products. These machineries will be available as manual, semi- or fully automatic systems. A polymeric composite products manufacturer will be able to also find sources for suppliers of ancillary equipment needed for any process when working closely with a machinery supplier to select the best system/systems for the proposed manufacturers. Another important aspect is finding sources for suitable mini versions for laboratory work.

| Supplier | Country | Processing systems |
|---|---|---|
| Hardy Smith Ltd. | India | Extrusion |
| Reifenhauser GmbH & Co | Germany | Extrusion |
| Harden Industries Ltd. | China | Extrusion |
| Davis-Standard, LLC | USA | Injection molding/extrusion |
| Coperion K-Tron | USA | Mixing/drying |
| Wuhan Plastics Machinery Ltd. | China | Composite polymers |
| Hennecke GmbH | Germany | PUR foaming |
| Karunanand Hydropneumatic Controls Ltd. | India | Compression molding |
| HAMRO International Co. Ltd. | Taiwan | Compression molding |

https://doi.org/10.1515/9783110656152-013

# Appendix C

The following table shows some of the popular manufacturers and suppliers of polymers. They will be able to advise processors about good sources for additives, dyes, pigments and others. Polymers are generally available as powders, liquids, granules or pellets, in either natural color or self-colored. Basic packs include 25 kg paper bags or larger 400–500 lb bulk packs, except for the polymers in liquid form. Countries shown are the main manufacturing sources but their products will be available from their agents/distributors in many countries.

| Supplier | Country | Products |
|---|---|---|
| Dow Corporation | USA | All |
| BASF | Germany | All – specialty EPS |
| Bayer AG | Germany | All – specialty PUR |
| Huntsman Corporation | Europe | Specialty PUR |
| ChemControl Limited | USA | PUR |
| Issac Industries Inc. | USA | PUR |
| Era Polymers Ltd. | Australia | Two component PUR systems |
| Biobased Technologies, LLC | USA | Specialty polymers |
| Union Carbide Limited | Canada | Polymers |

This table, compiled by the author, highlights a few major suppliers.

https://doi.org/10.1515/9783110656152-014

# Glossary

**Amorphous** Having no ordered arrangement. Polymers are amorphous when their chains are tangled up in any old way. Polymers are *not* amorphous when their chains are lined up in ordered crystals.

**Anion** An atom or molecule that has a negative electrical charge.

**Cation** An atom or molecule that has a positive electrical charge.

**Complex** Two or more molecules that are associated together by some type of interaction of electrons, other than a covalent bond.

**Copolymer** A polymer made from more than one type of monomer.

**Covalent bond** A joining of two atoms when the two share a pair of electrons.

**Cross-linking** Cross-linking occurs when individual polymer chains are linked together by covalent bonds to form one giant molecule.

**Crystal** A mass of molecules arranged in a neat and orderly fashion. In polymer crystal, the chains are lined up neatly like new pencils in a package. They are also bound together tightly by secondary interactions.

**Elastomer** Rubber. Hot shot scientists say a rubber or elastomer is any material that can be stretched many times its original length without breaking, *and* will snap back to its original size when it is released.

**Electrolyte** A molecule that separates into a cation and an anion when it is dissolved in a solvent, usually water. For example, NaCl (salt) separates into $Na^+$ and $Cl^-$ in water.

**Elongation** How long a sample is stretched when it is pulled? Elongation is usually expressed as the length after stretching divided by the original length.

**Emulsion** A mixture in which two immiscible substances, such as oil and water, stay mixed together thanks to a third substance called an *emulsifier*. The emulsifier is usually something like a soap, whose molecules have a water-soluble end and an organic-soluble end. The soap molecules form little balls called *micelles*, in which the water-soluble ends point out into the water, and the organic-soluble ends point into the inside of the ball. The oil is stabilized in the water by hiding in the center of the micelle. Thus, the water and oil stay mixed.

**Entropy** Disorder. Entropy is a measure of the disorder of a system.

**First-order transition** A thermal transition that involves both a latent heat and a change in the heat capacity of the material.

**Free radical** An atom or molecule which has at least one electron which is not paired with another electron.

**Gel** A cross-linked polymer which has absorbed a large amount of solvent. Cross-linked polymers usually swell a good deal when they absorb solvents.

**Gem diol** A diol in which both hydroxy groups are on the same carbon. Gem diols are unstable. Why are they called *gem* diols? It is short for *geminal*, which means "twins." It is related to the word *gemini*.

https://doi.org/10.1515/9783110656152-015

**Glass transition temperature** The temperature at which a polymer changes from hard and brittle to soft and pliable.

**Heat capacity** The amount of heat it takes to raise the temperature of 1 g of a material 1 °C.

**Hydrodynamic volume** The volume of a polymer coil when it is in solution. This can vary for a polymer depending on how well it interacts with the solvent, and the polymer's molecular weight.

**Hydrogen bond** A very strong attraction between a hydrogen atom which is attached to an electronegative atom, and an electronegative atom which is usually on another molecule. For example, the hydrogen atoms on one water molecule are very strongly attracted to the oxygen atoms on another water molecule.

**Ion** An atom or molecule that has a positive or a negative electrical charge.

**Latent heat** The heat given off or absorbed when a material melts or freezes or boils or condenses. For example, when ice is heated, once the temperature reaches 0 °C, its temperature will not increase until all the ice is melted. The ice has to absorb heat in order to melt. But even though it is absorbing heat, its temperature stays the same until all the ice has melted. The heat required to melt the ice is called the *latent* heat. The water will give off the same amount of latent heat when you freeze it.

**Le Chatlier's principle** This principle states that if a system is placed under stress, it will act so as to relieve the stress. Applied to chemical reactions, it means that if product or byproduct is removed from the system, the equilibrium will be upset, and the reaction will produce more products to make up for the loss. In polymerizations, this trick is used to make polymerization reactions reach high conversion.

**Ligand** An atom or group of atoms that are associated with a metal atom in a complex. Ligands may be neutral or they may be ions.

**Living polymerization** A polymerization reaction in which there is no termination and the polymer chains continue to grow as long as there are monomer molecules to add to the growing chain.

**Matrix** In a fiber-reinforced composite, the matrix is the material in which the fiber is embedded, the material that the fiber reinforces. It comes from a Latin word, which means "mother," interestingly enough.

**Modulus** The ability of a sample of a material to resist deformation. Modulus is usually expressed as the ratio of stress exerted on the sample to the amount of deformation. For example, tensile modulus is the ration of stress applied to the elongation which results from the stress.

**Monomer** A small molecule that may react chemically to link together with other molecules of the same type to form a large molecule called a polymer.

**Olefin metathesis** A reaction between two molecules containing carbon–carbon double bonds. In olefin metathesis, the double bond carbon atoms change partners, to create two new molecules, both containing carbon–carbon double bonds.

**Oligomer** A polymer whose molecular weight is too low to really be considered a polymer. Oligomers have molecular weights in hundreds, but polymers have molecular weights in thousands or higher.

**Plasticizer** A small molecule that is added to the polymer to lower its glass transition temperature.

**Random coil** The shape of a polymer molecule when it is in solution, and it is all tangled up in itself, instead of being stretched out in a line. The random coil only forms when the intermolecular forces between the polymer and the solvent are equal to the forces between the solvent molecules themselves and the forces between polymer chain segments.

**Ring-opening polymerization** A polymerization in which cyclic monomer is converted into a polymer that does not contain rings. The monomer rings are opened up and stretched out in the polymer chain, like this.

**Secondary interaction** Interaction between two atoms or molecules other than a covalent bond. Secondary interactions include hydrogen bonding, ionic interaction and dispersion forces.

**Second-order transition** A thermal transition that involves a change in heat capacity, but does not have a latent heat. The glass transition is a second-order transition.

**Soap** A molecule in which one end is polar and water-soluble and the other end is nonpolar and organic-soluble, such as sodium lauryl sulfate.

These form micelles in water, little balls in which the polar ends of the molecules point out into the water, and the nonpolar ends point inward, away from the water. Water-insoluble dirt can hide inside the micelle, so soapy water washes away dirt that plain water cannot.

**Strain** The amount of deformation a sample undergoes when one puts it under stress. Strain can be elongation, bending, compression or any other type of deformation.

**Strength** The amount of stress an object can receive before it breaks.

**Stress** The amount of force exerted on an object, divided by the cross-sectional area of the object. The cross-sectional area is the area of a cross section of the object, in a plane perpendicular to the direction of the force. Stress is usually expressed in units of force divided by area, such as $N/cm^2$.

**Termination** In a chain growth polymerization, termination is the reaction that causes the growing chain to stop growing. Termination reactions are reactions in which none of the products may react to make a polymer grow.

**Thermoplastic** A material that can be molded and shaped when it is heated.

**Thermal transition** A change that takes place in a material when you heat it or cool it, such as melting, crystallization or the glass transition.

**Thermoset** A hard and stiff cross-linked material. Thermosets are different from *thermoplastics*, which become moldable when heated. Thermosets are cross-linked, so they do not. Also, they are different from cross-linked *elastomers*. Thermosets are stiff and do not stretch the way elastomers do.

**Toughness** A measure of the ability of a sample to absorb mechanical energy without breaking, usually defined as the area underneath a stress–strain curve.

**Transesterification** A reaction between an ester and an alcohol in which the -O-R of the ester and the -O-R' group of the alcohol exchanges.

# Acknowledgments

In compiling this book, the author has carried out intense research in order to make a very comprehensive presentation. He is grateful to many industrial companies that have shared their technical know-how and also have supplied photos of their products.

Combined with the author's 45 plus years of hands-on industrial experience, the reader is assured of very valuable information on main aspects of processing polymers. Detailed information from actual practice of projects set up by the author in Sri Lanka, Canada and the Philippines is also provided for the benefit of small processors and entrepreneurs.

He is also grateful to the editors and staff of De Gruyter for the excellent support in preparation and editing to make this book possible.

Chris Defonseka

https://doi.org/10.1515/9783110656152-016

# Index

https://doi.org/10.1515/9783110656152-017

www.ingramcontent.com/pod-product-compliance
Lightning Source LLC
Chambersburg PA
CBHW081533220326
41598CB00036B/6421